Social Fuzziology

Studies in Fuzziness and Soft Computing

Editor-in-chief
Prof. Janusz Kacprzyk
Systems Research Institute
Polish Academy of Sciences
ul. Newelska 6
01-447 Warsaw, Poland
E-mail: kacprzyk@ibspan.waw.pl
http://www.springer.de/cgi-bin/search_book.pl?series=2941

Further volumes of this series can be found at our homepage.

Vol. 87. C. Bertoluzza, M. Á. Gil and D. A. Ralescu (Eds.)
Statistical Modeling, Analysis and Management of Fuzzy Data, 2002
ISBN 3-7908-1440-7

Vol. 88. R. P. Srivastava and T. J. Mock (Eds.)
Belief Functions in Business Decisions, 2002
ISBN 3-7908-1451-2

Vol. 89. B. Bouchon-Meunier, J. Gutiérrez-Ríos, L. Magdalena and R. R. Yager (Eds.)
Technologies for Constructing Intelligent Systems 1, 2002
ISBN 3-7908-1454-7

Vol. 90. B. Bouchon-Meunier, J. Gutiérrez-Ríos, L. Magdalena and R. R. Yager (Eds.)
Technologies for Constructing Intelligent Systems 2, 2002
ISBN 3-7908-1455-5

Vol. 91. J. J. Buckley, E. Eslami and T. Feuring
Fuzzy Mathematics in Economics and Engineering, 2002
ISBN 3-7908-1456-3

Vol. 92. P. P. Angelov
Evolving Rule-Based Models, 2002
ISBN 3-7908-1457-1

Vol. 93. V. V. Cross and T. A. Sudkamp
Similarity and Compatibility in Fuzzy Set Theory, 2002
ISBN 3-7908-1458-X

Vol. 94. M. MacCrimmon and P. Tillers (Eds.)
The Dynamics of Judicial Proof, 2002
ISBN 3-7908-1459-8

Vol. 95. T. Y. Lin, Y. Y. Yao and L. A. Zadeh (Eds.)
Data Mining, Rough Sets and Granular Computing, 2002
ISBN 3-7908-1461-X

Vol. 96. M. Schmitt, H.-N. Teodorescu, A. Jain, A. Jain, S. Jain and L. C. Jain (Eds.)
Computational Intelligence Processing in Medical Diagnosis, 2002
ISBN 3-7908-1463-6

Vol. 97. T. Calvo, G. Mayor and R. Mesiar (Eds.)
Aggregation Operators, 2002
ISBN 3-7908-1468-7

Vol. 98. L. C. Jain, Z. Chen and N. Ichalkaranje (Eds.)
Intelligent Agents and Their Applications, 2002
ISBN 3-7908-1469-5

Vol. 99. C. Huang and Y. Shi
Towards Efficient Fuzzy Information Processing, 2002
ISBN 3-7908-1475-X

Vol. 100. S.-H. Chen (Ed.)
Evolutionary Computation in Economics and Finance, 2002
ISBN 3-7908-1476-8

Vol. 101. S. J. Ovaska and L. M. Sztandera (Eds.)
Soft Computing in Industrial Electronics, 2002
ISBN 3-7908-1477-6

Vol. 102. B. Liu
Theory and Praxis of Uncertain Programming, 2002
ISBN 3-7908-1490-3

Vol. 103. N. Barnes and Z.-Q. Liu
Knowledge-Based Vision-Guided Robots, 2002
ISBN 3-7908-1494-6

Vol. 104. F. Rothlauf
Representations for Genetic and Evolutionary Algorithms, 2002
ISBN 3-7908-1496-2

Vol. 105. J. Segovia, P. S. Szczepaniak and M. Niedzwiedzinski (Eds.)
E-Commerce and Intelligent Methods, 2002
ISBN 3-7908-1499-7

Vol. 106. P. Matsakis and L. M. Sztandera (Eds.)
Applying Soft Computing in Defining Spatial Relations, 2002
ISBN 3-7908-1504-7

Vladimir Dimitrov
Bob Hodge

Social Fuzziology

Study of Fuzziness of Social Complexity

Physica-Verlag
A Springer-Verlag Company

Dr. Vladimir Dimitrov
Prof. Bob Hodge
University of Western Sydney
Locked Bag 1797
South Penrith DC
NSW 1797
Australia

ISSN 1434-9922
ISBN 3-7908-1506-3 Physica-Verlag Heidelberg New York

Library of Congress Cataloging-in-Publication Data applied for
Die Deutsche Bibliothek – CIP-Einheitsaufnahme
Dimitrov, Vladimir: Social fuzziology: study of fuzziness of social complexity / Vladimir Dimitrov; Bob Hodge . – Heidelberg; New York: Physica-Verl., 2002
(Studies in fuzziness and soft computing; Vol. 107)
ISBN 3-7908-1506-3

Physica-Verlag Heidelberg New York
a member of BertelsmannSpringer Science+Business Media GmbH

Printed in Germany

Hardcover Design: Erich Kirchner, Heidelberg

SPIN 10882593 88/2202-5 4 3 2 1 0 – Printed on acid-free paper

To our mothers

Prologue

In 1970 I started my journey into the social field of application of fuzzy sets with the design of a heuristic model for decision-making (Dimitrov, 1970); I then worked on developing fuzzy questionnaires for studying public opinion (Barnev, Dimitrov and Stanchev, 1974). My research interest was focused in the management of human systems (Dimitrov and Wechler, 1975) and their modelling (Dimitrov, 1977), exploration of the group choice problem and social self-organisation (Dimitrov, 1976a), application of fuzzy sets to economics (Dimitrov and Angelkov, 1976), medical diagnostics (Dimitrov, Stanchev and Driankov, 1976), water pollution control (Dimitrov, Driankov, Petrov, 1977), etc. This inspiring stage of research ended with the development of *non-formal theory of fuzzy management with applications* (Dimitrov, 1976) - a work for which in 1979 the expert council on mathematical cybernetics of the Academy of Sciences of the former Soviet Union conferred on me the degree "doctor of science". Resulting from this work was the realisation that the fuzzy management of human systems, that is, management which tolerates imprecisely formulated instructions, is more efficient than any rigid deterministic control. The famous social choice paradox of Arrow (Arrow, 1963) 'melts' when the individual preferences are expressed fuzzily. In this case, the fuzziness converges into two fuzzy values: one centred in the fuzzy set of alternatives considered as *significant* by the individuals and the other centred in the fuzzy set of *insignificant* alternatives. This type of convergence helps avoid the undesirable emergence of triads with 'vicious' preferences (such as: *a* preferred to *b* preferred to *c*, and then *c* preferred to *a*), as it transforms the initial multi-alternative choice into a choice between two fuzzy values, which cannot be entrapped in vicious triads.

Quite possibly, the concept of fuzzy management was unacceptable for the totalitarian regime at that time, so my degree was rejected by the Highest Certifying Commission of the Council of Ministers of the USSR. This decision was re-considered when Gorbachev started his *perestroika* and the DSc degree conferred on me by the Academy of Sciences in Moscow was officially re-confirmed in 1987. So, it took 8 years for the Soviet officials to accept the 'dangerous' results of the application of fuzzy sets to social reality. I had no problems publishing in the Western journals and proceedings of international congresses and conferences: Dimitrov, 1977a; Dimitrov and Kunchev, 1977b; Dimitrov and Driankova, 1977; Dimitrov and Driankov, 1977; Dimitrov, 1983; Dimitrov, 1989. My meetings with Prof. Zadeh - the founder of fuzzy sets - in

At the end of 1999, as an academic in the University of Western Sydney, Australia, I was ready to describe the main ideas of **fuzziology** as a *study of the sources, nature and dynamics of the fuzziness of the human knowing, understanding, and experiencing of reality.*

Fuzziology expands the exploratory field of social reality beyond the reach of the fuzzy logic. While fuzzy logic agrees that both A *and* non-A (where A can be an issue of concern, alternative, control value, etc.) are considered simultaneously true up to some degrees, the endeavour of fuzziology is to transcend (go beyond) the duality of both fuzziness *and* non-fuzziness in people's understanding. This endeavour reflects the ancient Vedic maxim: "*Do not reject anything! But do not remain with anything either! Go beyond*!"

By postulating that the self-referential processes of our understanding and knowing grow from within us and have the potential to transcend any limits and obstacles, the 'transcendent' logic of fuzziology serves as a powerful mind-energiser - stimulator of our creativity and catalyst for seeking mutual understanding and social harmony. What appears fuzzy or non-fuzzy in the context of some social issues, which we consider as important at a certain level of development of our understanding, dissolves or disappears when our consciousness expands. By moving beyond the current level of understanding and knowing, we open a new space of social possibilities, and thus 'transcend' both the fuzziness and non-fuzziness existing at the previous level.

The interesting and useful discussions on fuzziology, which I have with my colleague and friend - professor Bob Hodge (at the College of Arts, Education and Social Sciences of the University of Western Sydney), helped us to initiate a new fertile zone of inquiry which we named *social fuzziology*. So, here is our first joint book. I wanted the text of our first book not to be so long; in the ever-growing busyness of our every day lives we simply do not have much time for reading books with many pages, have we? A lot of ideas that I shared with Bob are not included in the book (you can have a look at our home page at http://www.uws.edu.au/vip/dimitrov/). If the reader - social researcher, engineer, philosopher, academic, student, explorer of the perennial wisdom and Vedic spirituality or anyone who likes to deal with ideas that challenge and irritate - finds this book interesting (and share this with us) we can soon be ready with our next book.

Vlad Dimitrov
University of Western Sydney

May 2002

Foreword

Fuzziology studies the fuzziness inherent in what we know about ourselves, the sources and nature of our experience, our thoughts and feelings, drives for understanding and urges to create and realise our potential. This kind of fuzziness is at the core of our existence, at the essence of our humanness. It affects any field of human activity, be it mathematical study of fuzzy equations and fuzzy integrals; engineering design and implementation of fuzzy logic-based methodologies; fuzzy control systems or fuzzy robots.

Social fuzziology investigates the role of fuzziness in understanding the dynamic complexity of human existence in the social world. It is a study of the nexus between the complex demands of life - individual and social - and the fuzziness of thinking.

Since human evolution over 2 billion years has seen the co-evolution of social complexity with human language and thought, it is likely that the fuzziness of language and thought is especially intimately formed by the demands of social complexity, just as social complexity is sustained by the inherent fuzziness of language and thought.

Social fuzziology is not simply one field of application of fuzziology. Given the initial state of the development of fuzziology, social fuzziology needs to develop hand in hand with fuzziology, each helping to advance the other.

The scope of the book is reflected in the background of the authors, one from science and mathematics, the other from semiotics and social sciences. This also determines how we have written, and the audience we hope to reach. That audience, we hope, will include two groups who normally never read the same books, who live in two different 'cultures', science and humanities, which have grown too far apart and need to be reintegrated. We have tried to make the text equally accessible to both kinds of reader. We hope each will tolerate kinds of arguing that may seem unfamiliar from their disciplinary perspective, or explanations of matters too well known (to experts in the field) to need saying.

There are many positive benefits to both sides for such efforts. As such notable scientists as Einstein and Linus Pauling amongst many others have proclaimed, scientists and engineers have so much impact on society they cannot afford to be ignorant or indifferent about the great issues of their day. Conversely, the social sciences today are mostly based on a particular understanding of science which lacks insights of social fuzziology. They need the kind of renewal social fuzziology could offer, if they are to develop better tools for understanding the escalating complexities of contemporary society.

Social fuzziology is so new it has only just acquired a name, yet fuzziness is too important and pervasive for it never to have been noticed or reflected on by any earlier thinkers or traditions. Indeed, we believe that the opposite is the case. Ancient traditions of wisdom often incorporate a profound recognition of the nature and role of fuzziness, under other terms or in other systems of thought. Many modern thinkers have also reflected in interesting and valid ways on fuzziness. To those who would demand that we define fuzziology or social fuzziology in exclusive terms, distinct from each other or from other fields of thought, we say: such a definition would contradict a fundamental principle of fuzziology itself.

In the book we explore many issues of fuzziology and social fuzziology. Here we put forward the following propositions:

1. What we experience, learn and know can be considered as a mapping of the all-pervading matter of the universe onto the 'transcendental' space of human ideas.

The universe is a material expression of all-pervading dynamics. The interplay of energies, forces, forms, substances produces self-organizing drives of various kinds and magnitudes. They are responsible for the organization of matter into material reality, which we try to perceive and study through our senses coordinated by our mind, while using instruments which aim to magnify the ability of our senses.

The inner space of our thoughts, emotions, feelings, beliefs, aspirations, dreams, etc. is a product of the same universal dynamics, and preserves its essentially dynamical character. Thoughts and emotions are in constant movement - they emerge, interact, disappear, and emerge again. They penetrate into one another, weave into complex concepts and feelings, habits and stories. Their 'boundaries' are not only fluid and permeable, they expand, shrink, combine, 'fractalise', 'bifurcate', etc. In this sense we consider them fuzzy emanations of our inner space. Since the whole process of human knowing is based on these emanations, our knowledge is inherently fuzzy.

2. We human beings are the main source of the fuzziness (uncertainty, vagueness, ambiguity, indeterminacy, unclearness, indistinctness, obscurity) that permeates our experience and being, our understanding and knowing, our 'languaging' and acting.

Being a product of the self-organizing and evolving dynamics of nature, and at the same time, a factor in their changes, it is difficult for us to grasp what is the engine of these dynamics, where it is located and how it works. We try to answer these questions by studying the world that surrounds us, and while doing so we separate ourselves from the world. The mind, which seeks to understand the world, remains here, centred in our brains and senses, while the world we observe and study spreads over there, in the universe.

We think that the world evolves in a way different from how we evolve, having its own engine different from the engine supporting our lives. Following this logic, our mind (constantly involved in analysing, classifying and separating) takes the next step: it supposes that the source of all difficulties which emerge on the way to understand the world are in the world alone, in its variety, complexity and vastness. What a relief is this for the individual mind! Instead of trying to understand itself, to see what impedes its capacity to concentrate, create and evolve, the individual thinker hurries to blame the external world and its unsurpassable problems, forgetting that it is the thinker who has created and continue to create the most severe problems in the world.

3. There is no fuzziness located outside human nature, in a world separated from us. Such a separated-from-us world does not exist. We and the world are forever connected through the umbilical cord of the all-pervading common dynamics - energies, forces, substances and forms. They have created us in the process of their interactions. They support our existence and vitally affect our lives, but we also contribute in the ways they manifest through our evolving nature.

Nature does not select a special kind of dynamics to support our existence, and another kind to support the existence of the rest of the world. The living cells of our finite bodies and the constellations of our infinite universe share one and the same *energy field* - a field that sustains the life and evolution of the animated nature and stimulates changes and transformations of inanimate forms. The ways we experience and live in this field, the ways we interpret and explain its numberless manifestations, the ways we make sense of these manifestations, appear *fuzzy* to our minds.

What we see and sense as existing outside of our bodies affect our inner experience and knowledge, and vice versa: our inner experience and knowledge crucially affect how we see the world; so the separation between inner and outer is artificial. Moreover, such a separation gives rise to delusion and feeds our ignorance. The fuzziness of our experience, of our thoughts, feelings, words and actions does not inhibit our urge to understand the enigmas and paradoxes of existence; on the contrary, it is only because of this kind of fuzziness that our urge to know and our search for truth continue forever. If we were able to know everything, there would be no need to strive for knowledge and truth.

4. Fuzziness is not in a steady state, but moves, expands and shrinks, bifurcates or 'explodes' into many streams. The broader and deeper the human inquiry, the more saturated with overlapping or interwoven streams of ever-moving and changing fuzzy ideas and stories, hypotheses and theories, emotions and feelings, beliefs and expectations, aspirations and dreams.

The *dynamics of fuzziness* never move in a straight-line, from the unknown to the known. One can imagine them in a *vortex-like* motion, like the dynamics of a whirlpool or tornado. What we assume as known today becomes fuzzy tomorrow only to appear clear after a while, and again fuzzy a moment later. The natural ebbs and tides of fuzziness continue forever. In the similar way as the vorticity of a whirlpool or a tornado produces forces of significant magnitude, the 'vorticity' of fuzziness produces and sustains the urge to know and reveal the true nature of reality. This urge leads to the emergence of insights, raptures of inspiration, enlightenment and wisdom.

5. The larger the 'vorticity' of the fuzziness of our experience and knowledge, that is, the richer and more impregnated-with-dynamics their fields, the greater the chance of epiphany - a sudden intuitive sparkle in our thinking, a spontaneous perception of the essential nature of reality, an illuminating discovery of truth.

Fuzziology refers to such epiphany as 'bootstrapping' of fuzziness; the fuzziness in one's understanding pulls itself by its own bootstraps and moves to another level. The levels of fuzziness correspond to the levels of our capacity for understanding, to the levels of our consciousness. To say that the fuzziness has moved to another level means that the process of our understanding has moved to another level also, and what was fuzzy and incomprehensible for the mind at the level, from where the fuzziness has pulled itself, has become clear and comprehensible. Of course, this does not mean that there is no more fuzziness in our understanding, that we have won the battle with the fuzziness and succeeded to extinguish it once and for all. The fuzziness is 'alive' - full of vigour and potential to become denser and to expand wider, but its dynamics are 'whirling' at another level. One can call the new level 'higher' or 'deeper', it does not matter; what matters is that one's understanding has become deeper, that one's consciousness has been expanded to a higher level, that the limitations, which fuzziness used to impose on one's thinking at the previous level, have been transcended. The inquiring mind will soon encounter the limitations that the fuzziness will bring with at the new level of its evolving dynamics, so that to challenge the mind to explore it further and make it move again.

Social fuzziology is given a twofold role in the turbulent times of the 21st century: to recognise the fuzziness of human understanding while negotiating complexity with integrity, truth and an ethical dimension. For all its difficulties, social fuzziology must fulfil this double task.

A book like this will have many messages for many readers. There are five messages of social fuzziology we especially want to emphasise here.

The fuzziness of human knowing is not a burden, but welcome. It is through fuzziness that we can understand society better and more constructively: a better guide than any dogma, whether ideological, religious, or scientific.

When fuzziness is seen as a problem, 'the cure is in the poison'; the solution may be found in the fuzziness itself, which can allow a qualitative leap, into a higher level of insight or understanding.

In social situations dominated by the exercise of power (inequality, injustice, exploitation), both the lack of fuzziness (crisp authoritarian logics, dogmatic thinking) and the imposed fuzziness (delusive, manipulative fuzziness) tend to cause situations to deteriorate, leading to suppression of people's creativity, stress and lower levels of social well-being.

Social fuzziology can ameliorate many social situations by showing how to fuzzify divisions and boundaries that mark and deform every society (such as racism, discrimination, prejudice, exploitation on grounds of class, ethnicity, age or belief) the same time as it identifies boundaries that are more flexible and effective in sustaining the diversity and variety of human societies.

In situations of high intensity and high levels of conflict, fuzziness takes the form of contradictions (bifurcations, polarization, fanaticism, irrationality) in which crisp and fuzzy thinking alike may prove ineffective and counterproductive. In such situations social fuzziology needs to work with a different kind of logic, transcendent logic, in order to discover unforeseen possibilities of resolution and harmony.

Vlad Dimitrov and Bob Hodge
University of Western Sydney

May 2002

Table of Contents

1 Introduction to Social Fuzziology *1*

1.1 The Basic Postulate of Fuzziology *1*

1.2 The Approach of Fuzziology *3*

1.2.1 Dynamic Character of Fuzziness *3*
1.2.2 The Bootstrapping Algorithm *4*
1.2.3 Paradox of Fuzziology *5*
1.2.4 Example *7*

1.3 The Project of Social Fuzziology *10*

1.3.1 Science and Art of Social Fuzziology *10*
1.3.2 Awareness of Life as it Unfolds *11*

1.4 Socrates' Wisdom at the Origin of Fuzziology *12*

1.4.1 Meno's Paradox *12*
1.4.2 Socrates' Approach *14*
1.4.3 Maieutic Inquiry *15*
1.4.3.1 Conditions Enhancing Maieutic Inquiry *17*
1.4.3.2 Maieutic Way: From Knowledge to Wisdom *18*

1. 5 Principles of Fuzziology: A Social Perspective *20*

1.5.1 Principle of Incompatibility *20*
1.5.2 Principle of Connectivity-in-Dynamics *21*
1.5.3 Principle of Fractality *21*

1.6 Impossibility Theorems *22*

1.6.1 First Impossibility Theorem *22*
1.6.2 Second Impossibility Theorem *24*

1.7 Social Nature of Fuzziness *26*

1.7.1 The Double Face of Fuzziness *26*

1.7.2 Fuzziness and Power 27

1.8 Human-centred Sources of Fuzziness 29

1.8.1 Mind as a Source of Fuzziness 29
1.8.2 Ever-emerging Desires 30
1.8.3 'Golden Rules' of Fuzziology 31
1.8.4 Danger of Dogmatising Fuzziness 32
1.8.5 The 'Drowning-Man' Paradox 34

1.9 Transcending Duality and Non-duality 36

1.9.1 Polarisation 36
1.9.2 'Transcendent' Logic of Social Fuzziology 38

1.10 Consciousness Resonance 39

1.10.1 The Possibility Theorem 39
1.10.2 Journey Beyond Limitations of Fuzziness 40
1.10.3 Activating Consciousness Resonance 41
1.10.4 The 'Bootstrapping' Theorem 42
1.10.5 Resonance at Social Level 43

2 Bridging the Study of Complexity with Social Fuzziology 47

2.1 Complementarity and Mutual Enrichment 48

2.2 'Three Body Problem' in Fuzziology 50

2.2.1 KAM Theorem 50
2.2.2 Dynamics of Fuzziness 51

2.3 Chaotic Attractors of Meaning 52

2.3.1 Emergence of New Meanings 53
2.3.2 Meanings of Dominance and Suppression 55
2.3.3 'Optimal' Degree of Fuzziness 56
2.3.4 Symmetry Breaking and 'Trunsduction' 57

2.4 Key Propositions in Social Fuzziology: A Complexity Perspective 58

2.4.1 Interwovenness and Unpredictability *58*
2.4.2 Dealing with Complex Causality *59*
2.4.3 'Fractality' and Wholeness *61*
2.4.4 Far-from-Equilibrium Criticality *62*
2.4.5 Self-organization and Vorticity *64*
2.4.6 Harmonious Co-evolution *66*

2.5 Fuzzy Logic and Fuzziology *67*

2.5.1 Fuzziness and 'Softness' *68*
2.5.2 Role of Fuzziology in Engineering Practice *69*

2.6 Fuzziology and Creativity *71*

2.6.1 Einstein's View on Creativity *72*
2.6.2 Inspiration *73*
2.6.3 The Zen of Creativity *74*

3 Understanding Fuzziness of Ourselves *77*

3.1 Paradox of Mind *77*

3.2 Journey towards Understanding Ourselves *78*

3.3 Human Experiential Space *80*

3.3.1 Characteristics of Human Experiential Space *81*
3.3.2 Comparison between Experiential Fuzziness and Fuzziness of Thinking *82*
3.3.3 'Strange' Attractions in Human Experience *84*
3.3.4 Understanding The Thinker *85*

3.4 Maieutic Inquiry into the Essence of Human Dynamics *87*

3.4.1 Existential Whirlpool: The Approach of Fuzziology *87*
3.4.2 Timeless Essence: The Vedic Approach *89*
3.4.3 On Practical Realisation of the Vedic Approach *91*
3.4.4 Compatibility between the Vedic Approach and the Approach of Fuzziology *93*

3.4.5 The Most Difficult Maieutic Question 94

3.5 Empowering the Thinker 95

4 Understanding Fuzziness of Society 99

4.1 Infinite Life of Society versus Finite Lives of Individuals 99

4.2 Collective Mind: A Pile of Fragmented Knowledge 100

4.3 From Understanding Ourselves to Understanding Society 101

4.4 Lack of Wisdom in Society 103

4.5 Towards Establishment of Global Empire 105

4.5.1. Hegemonic Ideology 106
4.5.2 Phenomenal Brainwash 107
4.5.3 False Identities 108
4.5.4 Towards Establishment of Global Control 110
4.5.4.1 System of Fixed Identities 110
4.5.4.2 Control of Human Identities 111
4.5.4.3 Forces of Separation 112

4.6 Education and Learning in Society 113

4.6.1 Metabolism of 'Borrowed' Knowledge 114
4.6.2 Learning in Captivity of the Global Order 116

4.7 The Health Illusion 117

4.8 Dealing with Hard Negotiation 120

4.8.1 Four Typical 'Hard' Situations 121
4.8.2 Pursuit of Consensus: Practical Aspects 124
4.8.3 Transcending Paradoxical Situations 126
4.8.4 The Incoherence Principle 126

4.9 Conclusion 128

5 Case Studies: Understanding Fuzzy Social Categories 131

5.1 What is 'Australian'? 131

5.2 Use and Abuse of Fuzziness in Social Life 133

5.3 Fuzzy Borders in the Globalising World 137

5.4 'Mexican Americans', 'Chicanos' and Santana 139

6 Fuzziness of the West and the East 143

6.1 Fuzzy Dionysos 144

6.2 Philosophy of Flux 145

6.2.1 Zeno 145
6.2.2 Heracleitus 147
6.2.3 The Heretics 148

6.3 Other Masters of Fuzziology from the West 149

6.3.1 Aristotle 149
6.3.2 Hegel 151

6.4 Myths of Fuzziness 152

6.4.1 Wisdom of the Ancient Myths 152

6.4.1.1 Procrustean Bed 152
6.4.1.2 Gordian Knot 153
6.4.1.3 Achilles Heel 154
6.4.1.4 Pandora's Box 155

6.5 The Fuzzy Wisdom of Tao 156

6.6 Aboriginal Wisdom 160
6.7 Using Fuzziness of the Ancient Wisdom 163

6.7.1 Bridging Myths and Reality *164*
6.7.2 Example *165*

7 Key Terms in the Language of Fuzziology *167*

7.1 Fuzziness Defined *169*

7.2 The Logos of Logic *171*

7.3 Transformations *172*

7.4 Transcendence *174*

7.5 Understanding *176*

7.6 The Social *177*

7.7 Conclusion *178*

References *181*

1 Introduction to Social Fuzziology

I do not insist that my argument is right in all other respects, but I would contend at all costs both in word and deed as far as I could that we will be better men, braver and less idle, if we believe that one must search for the things that one does not know, rather than if we believe that it is not possible to find out what we do not know and that we must not look for it.

Socrates

1.1 The Basic Postulate of Fuzziology

In a broad sense, fuzziness is the opposite of precision. Everything that cannot be defined precisely (that is, according to some broadly accepted criteria or norms of precision) and everything that has no clearly described boundaries in space or time is considered a bearer of fuzziness. In a narrow sense, fuzziness relates to the definition of fuzzy sets as proposed by Zadeh (Zadeh, 1965): sets, the belongingness to which is measured by a membership function whose values are between 1 (full belongingness) and 0 (non-belongingness).

Mathematics has developed powerful tools for studying and dealing with the fuzziness - uncertainty, imprecision, vagueness - that researchers encounter when accumulating facts in different fields of their inquiry. Advanced methods of probability theory and mathematical statistics, probabilistic reasoning and Bayesian networks, rough sets and fuzzy logic help researchers not only describe and explain fuzziness, but also reduce it and, if possible, eliminate from their experimental findings, theoretical statements and practical solutions. Science always strives for precise, valid and reliable results; so does any purposeful action, any reason-based activity.

Fuzziology is not another mathematical study of fuzziness, and social fuzziology is not another field of application of fuzzy logic.

It is fuzziness that is in the focus of fuzziology - fuzziness inherent in what we know about ourselves, about the sources and nature of our experience, of our thoughts and feelings, drives for understanding and urges to create and realise our potential. This kind of fuzziness is at the core of our existence, at the essence of our humanness; therefore, it affects any field of human activity.

Fuzziness studied by fuzziology is not 'over there', not in an outer world separated from us, but in the inner world of our own experience, in the 'swarm' of our thoughts and ideas, emotions and feelings, beliefs and dreams. We see and understand as much from the outer world - the world in which we live - as we have already developed inside us while learning how to enrich our experience, hone our awareness, expand our consciousness and strengthen our capacity to sense, think, create and know.

The basic postulate of fuzziology is simple: *our understanding and knowing grow from within us and cannot be implanted or imposed from without.* Human understanding and knowing are self-organising processes; and any self-organising process in nature works from inside out. The universe expands due to forces that emerge from inside of its whirling dynamics. Every single seed grows from inside when the outer conditions do not impede but stimulate this growth; so does our understanding. It expands and grows from inside following the inner urge to know when there are conditions in the external world nourishing this urge and facilitating its realization. So, there is a role for human society to play - not to impede our inner drive for wisdom, but to encourage its outward fulfilment.

Our perceptions of reality, our experience of the events of life, our thinking and feeling, understanding and knowing are deeply rooted in the life of society and its development; we are simultaneously creators and products of society. The processes of our understanding and knowing are social in their origin and nature - fuzziness is imbedded in them. Therefore we refer to fuzziology as *social fuzziology.*

1.2 The Approach of Fuzziology

1.2.1 Dynamic Character of Fuzziness

Fuzziness is inherent in our perception of reality and in every kind of activity based on this perception, such as experiencing and making sense of the events of life, feeling and responding emotionally (while involved in communication with one another, with nature and with ourselves), thinking and speaking, learning and understanding, knowing, acting and creating.

When developing its approach to study fuzziness, fuzziology acknowledges its dynamic character and makes an emphasis on the following four points:

- Fuzziness has its sources and supporters, causes and effects, activators which cause it to increase and become denser and thicker, inhibitors which cause it to decrease and become rarer and thinner, exposers which allow it to become easily recognisable and obscurers which allow it to be hidden and difficult to disclose.

- Fuzziness has its own dynamics - forces and energies which make it move, change, evolve and transform, and its own carriers that are either immaterial like thoughts, ideas, feelings, emotions, longings, beliefs, dreams, aspirations, energy fields and spaces, or embodied in concrete human actions, in specific non-animated and animated forms, in discernible experiential events, in various kinds of signs and omens, phenomena and processes, human-created products and machines.

- Fuzziness is able to self-organise into dynamic patterns with boundaries that can become rigid and hard-to-surpass or soft and easy-to-permeate, to form attractors or repellers in the experiential or mental space of the individuals, to structure itself into layers (levels) going deeper into one's thoughts and feelings, or into whirlpools (vortices) of mental, emotional or spiritual energy producing creative forces - powerful individual urges and drives - which enable fuzziness to transcend the boundaries of its dynamic patterns and move from one level to another, from one attractor to another, from an individual to another.

- Fuzziness can never be fully eliminated from the human perception of reality and experience of life - from our thinking and feeling, from our understanding and knowing.

The *fuzziness* and *uncertainty* are identical in their meaning, if uncertainty is considered as embedded in human perception of reality. If uncertainty is seen as something outside the human ability to perceive, to experience, to understand and to know, that is, as something which exists 'over-there', in the 'objective' world that surrounds us, then uncertainty has another connotation than fuzziness. Fuzziness is a human characteristic, and not a characteristic intrinsic to an external object. Our knowledge about an external object can be fuzzy (vague, uncertain, ambiguous, obscure), but the object by itself has nothing fuzzy in its existence. The object is what the natural or human-created dynamics - forces, energies, substances and forms, which act upon it and express through its current appearance - have made it.

At the moment we consciously direct our attention towards an external object, the object 'enters' into the realm of our fuzziness - the fuzziness of our perception: experiencing, feeling, thinking, understanding, knowing, acting. We call this operation of including an external object into our fuzziness *interiorisation*, in analogy with Bakhtin's operation under the same name, proposed to define the process whereby each participant in a common dialogue adopts the story which emerges from this dialogue as his or her own (Morson and Emerson, 1989). After the interiorisation, the subject (the perceiver, the 'experiencer', the thinker, the knower, the actor) and the object (of perception, experience, understanding, knowing, acting) are no longer separated; they are linked through the dynamics of the subject's fuzziness.

1.2.2 The Bootstrapping Algorithm

The approach of fuzziology is entirely centred in the self-referential nature of the process of human understanding.

For us to understand an object (a phenomenon, a process, an experiential event, ourselves, society) means to go beyond the limits of our own fuzziness in relation to what we understand and know about this object. But in order to move beyond the fuzziness of our understanding, the only tool we can use is again our own understanding which carries the same fuzziness. So the process of understanding is a kind of realisation of a bootstrapping algorithm in the human mind, that is, seeding or facilitating emergence of conditions which helps fuzziness to *pull itself by its own bootstraps* and moves from one level (of one's understanding of reality) to another. The realisation of such bootstrapping algorithm becomes possible because the fuzziness is *dynamic* - it moves: shrinks and expands, accelerates and slows, hardens or softens, transforms and transcends its own dynamic patterns.

By studying our own fuzziness - its dynamic nature, sources, causes and factors which effect its motion, we are able to succeed in the activating of bootstrapping algorithms and help fuzziness transcend itself.

'Levels of fuzziness' correspond to the levels of our capacity for understanding, to the levels of development (growth) of our consciousness. To say that the fuzziness has moved to another level means that the process of our understanding has moved to another level also, and what was fuzzy and incomprehensible for the mind at the level, from where the fuzziness has pulled itself, has become clear and comprehensible. Of course, this does not mean that there is no more fuzziness in our understanding, that we have won the battle with the fuzziness and succeeded to extinguish it once and for all. Not at all!

Fuzziness is still 'alive' on each new level - full of vigour and potential to become denser and to expand wider. One can call the new level 'higher' or 'deeper', it does not matter; what matters is that one's understanding has become deeper, that one's consciousness has been expanded, that the limitations which fuzziness used to impose on the process of thinking at one level of one's understanding have been transcended. The inquiring mind will soon encounter the limitations that fuzziness brings with to the new level of its evolving dynamics; these limitations challenge the human mind to explore the fuzziness further and in this way cause it to move again.

1.2.3 Paradox of Fuzziology

What is important in applying the approach of fuzziology is that we do not need to fight with the fuzziness of our understanding in order to eliminate it. To eliminate fuzziness would be equivalent to stop developing our ability to perceive, experience, think, feel, understand, know and act, as fuzziness is inseparable from each and all of these vital processes.

The same motivation and urges which support the self-organisation of the human consciousness also support the self-organisation of the fuzziness - its ability to expand, shrink or 'pull itself by its own bootstraps'. The dynamics of the fuzziness inherent in one's understanding are, at the same time, dynamics of this very process of understanding, as understanding means nothing but overcoming (going beyond, transcending) the limitations of fuzziness embedded in this understanding, in its motion, changes and evolution.

How easy it would be, if it were possible to separate the fuzziness from the process of understanding, to isolate it and then either to eliminate or keep it in captivity, while victoriously moving outside its boundaries. Unfortunately, this is impossible! Fuzziness permeates the whole process of one's understanding and not only this, it permeates one's whole life, experience and consciousness.

The more we try to push fuzziness into only one region in our mental space - the space of our thoughts and ideas - or in our experiential space - the space where the trajectories of our lives unfold - the wider and denser its unexpected emergence in other regions.

When we create (seed, facilitate) conditions to energise and strengthen - broaden and deepen - the process of one's understanding, we simultaneously create conditions which energise the fuzziness dissolved in this process. Here lays the greatest *paradox of fuzziology,* no matter whether it is focused on studying the fuzziness of a single individual or the fuzziness of the society as a whole.

The higher the impetus to grow and evolve in consciousness, the more vigorous the expression of the fuzziness inherent in this growth and evolvement.

This paradox propels the development and application of **the approach of fuzziology** – an approach of:

- exploring the sources, nature, dynamics, causes and effects of the human fuzziness;

- not fighting with fuzziness, but trying to grasp its self-organising (bootstrapping) dynamics and to 'nudge gently' from within, in an almost unnoticeable manner. Such 'gentleness' and 'secrecy' is necessary in order not to provoke emergence of undesirable psychological reactions associated with resistance to changes in the human mind, as these are inevitable if there are well-established dynamic patterns of fuzziness in one's understanding and knowing (unfortunately, such patterns are always present in the human mind).

- activating the realisation of bootstrapping algorithms in human understanding by stimulating emergence of conditions whih help fuzziness 'pull itself by its own bootstraps', withdraw its limitations from a certain level of development of one's capacity to think, feel and experience, and then move to another level.

The above-formulated paradox puts emphasis on the significance of the practical realisation of the approach of fuzziology; each step in expanding its field of realisation has a greater value, as it deals with fuzziness of more potent nature.

At the same time, the paradox acts in favour of increasing the applicative power of the approach of fuzziology: the more 'virile' the fuzziness, the greater its capacity to transcend itself. This is of prime importance for the evolution of human thinking - for deepening of our understanding and expanding of our consciousness.

The paradox of fuzziology requires a high level of alertness at every stage of development of our consciousness to avoid absolutizing of what is considered known. According to the paradox of fuzziology, one can expect that the higher the level of consciousness (that is, the wiser the individual), the easier the fuzziness can pull itself from that level, and yet it is clear that efforts need to be applied and conscious actions to be undertaken for this to happen. With no efforts applied and no actions undertaken, fuzziness can never be made move, no matter how high its self-organising potential is.

1.2.4 Example

Let us apply the approach of fuzziology to explore the nature and sources of the fuzziness around the motives for the US to initiate a war against Afghanistan in November 2001. Officially, the war was characterised as a 'war against terrorism', with an immediate aim to capture Osama bin Laden and destroy his terrorist organisation al Qaida.

At the first level of the fuzziness around the motives for this war, we were prepared to almost immediately justify the war as an act of revenge (in the spirit of the rule: 'an eye for an eye and a tooth for a tooth') against the terrorist attacks that killed more than three thousands people in New York and Washington.

Now, when the war is over and more facts and evidences are accumulated to strengthen the capacity of our understanding, we ask ourselves: what this war really was aiming at? As any, the war caused the death of many people, not only soldiers but innocent people also, in Afghanistan - one of the world's poorest country, where more than 8 millions people have been forced from their homes by draught and civil war. The war put an end of the Taliban regime and a new interim government was formed. Osama bin Laden was not captured and the world terrorism was not destroyed, as the war had absolutely nothing to do with the removal of the deep social and economic roots and sources of the terrorism, nourished by today's monstrous power differential. The huge economic contrasts in society, the unsurpassable gaps of the ever-growing social inequalities and the inability of poor countries and marginalised social and ethnic minorities to resist the devastating effects of the global capitalist establishment on their economies, culture, urges for self-expression and independence are among the strongest catalysts of the terrorism all around the world - in the South and North America, in the Western Europe and Russia, in the Middle-East and Africa.

It was a 'war on terror' at superficial level only - war concerned with the effects and not with the causes of the terrorism. The overthrow of the Taliban government did not contribute much toward the weakening of the global terrorist network or even toward making the world order more democratic. There are many countries with political rulers even worse than the Talibans. The social policy of the government of Saudi Arabia - a country considered a friend of the US - differs little from the policy of the Taliban leaders, and yet the US has no intention to overthrow the regime in the Saudi Arabia. Arguably, the oil of that country has a much greater significance for the US than the suppression of the human rights there, and in particular the rights of the women.

When the fuzziness of our understanding moves and *self-organises*, based on more real-life material and contemplation, we may recall that in Afghanistan during the 1970s, with the support of Saudi Arabia and Pakistan - countries that could hardly be called democratic - Washington encouraged the creation of Islamic detachments recruited in the Arab-Islamic world and made up of what were called 'freedom fighters' at that time. As is well known now, this was the environment in which the CIA enlisted and trained Osama Bin Laden, so the latter was prepared as terrorist entirely by the US (Ramonet, 2001).

At the next level of the evolving dynamics of the fuzziness in our understanding of the war in Afghanistan, we may bring into consideration the previous attacks of the US on the Muslim world, such like Clinton's bombing of the Sudan without credible pretext - bombing that destroyed half its pharmaceutical supplies and probably killed tens of thousands of people (no one knows, because any inquiry at the UN into that event was blocked by the US); attacks on Iraq and Libya, which caused the death of innocent people, in order to 'give a lesson' to the governments of these countries (according to the official explanations). The US is the key military supporter to the Israeli occupation of the Palestinian territory and to the suppression of the Palestinian people's endeavour for freedom and independence. The US embargo against Iraq continues, causing the deaths of thousands of innocent civilians while preserving the regime in power. Before 1989, in the years of the could war, the US was actively involved in a crusade against communism in the Islamic world. Sometimes that resulted in mass extermination: thousands of communists killed in Iran; almost 1m communists killed in Indonesia. So there are certainly reasons for the people of the Islamic world to hold hostile feelings towards the US politics.

The *bootstrapping algorithm* of the fuzziness in our understanding of the war in Afghanistan is significantly facilitated when one becomes aware of the gigantic appetite of the Texas petrol companies - the most powerful supporters of the President G. W. Bush administration - for the resources of oil located in the Caspian region. The huge oil treasure of that region its extremely high qualities irresistibly keeps attracting the economic interests of the petrol magnates in the US for a long time. The US government cannot help but acts as an armed militia for the richest corporations in this country, for without the support of these corporations US Presidents would have no chance to be 'freely' elected or to make their decisions work (the 2001 election campaign of the president Bush cost 200 millions dollars!). The thirst of the US petrol giants for an ever-greater economic power might have had the ultimate word in the declaration of the first war conflict in the new millennium. Everything else - all that meaningless fuzzy charisma of the official speeches about saving the world from terrorism, bringing the terrorists to justice, defending freedom and democracy, helping the people in the US and the poor countries in the world - appears to be nothing but a brainwash to 'fuzzify' and delude people's understanding.

The war in Afghanistan had almost zero effects on preventing the emergence of new acts of terrorism and new Osmana-Bin-Ladens in the world, but it has perfectly served the insatiable thirst for military and economic power of the producers of advanced military tools for killing and for the petrol giants in the US.

It needs to be noted that the application of the approach of fuzziology for understanding a particular event, phenomenon or process has never an absolute ending; it is ready to take into account consideration of new facts and developments as they appear and unfold.

1.3 The Project of Social Fuzziology

Social fuzziology explores fuzziness in our understanding of society and ourselves as its creators and products at the same time. 'Society' is not simply an object out there, to be understood well or poorly through theories and categories. Society is a co-constructed object, a collective but not necessarily consensual set of meanings, expectations, roles and duties constraining the actions of self and others, as agents and affected participants, on-goingly created by interactions over time that are mediated through kinds of communication, which always appears to some extent fuzzy to the human mind. 'Society' in this sense exists as a network of fuzzy images, sustained by human thought and action. Society itself does not exist outside these processes. A group of human bodies is not a society, and without reference to these processes we cannot understand what any society does, as an entity or as an effect on actions and behaviours of individuals who compose it.

1.3.1 Science and Art of Social Fuzziology

The constant interplay of human dynamics at the three major scales of their manifestation: individual (intrapersonal dynamics), social (interpersonal dynamics) and existential (universal dynamics), results in the emergence of spinning webs and whirlpools of social interactions, which constantly reproduce forces and energies in order to strengthen or weaken the self-propelling capacity of these dynamics. There are so many intricately interwoven factors and conditions engaged in the realisation of this capacity of self-propulsion, that it is nonsensical to look for or to apply precise descriptions and definitions when explaining or dealing with their infinite, in number and diversity, embodiments.

Fuzziness has a crucial presence in our knowledge about ourselves and society. It is present or denied to different degrees in different theories of society (sociology, politics, history etc.) so the recognition of the role of fuzziness in any theory becomes a part of the evaluation of its adequacy. It is also present in the minds of social agents - politicians, prophets, advertisers, heads of media organisations, managers of companies and corporations, heads of academic departments, and also in members of the public, concerned citizens and activists, parents, children, lovers, friends. Everyone develops and internalises one or more theories of politics and society to live by. Everyone is a practical social theorist, and the theories people act on are always, to a greater or less extent, devices used to deal with the fuzziness of human knowing as it constitutes society and social processes.

Social fuzziology is a science concerned with social phenomena, so it can be classified as one of the social sciences. In theory as in practice, this could be a damagingly limited way of treating it, if it means that scientists and engineers come to think that it does not concern their core expertise. Engineering and scientific decisions are always taken in the light of assumptions and understandings, explicit and implicit, well-grounded or not, about the social, natural and constructed world which cannot be isolated from science or engineering problems, and the consequences of particular solutions.

Social fuzziology is also a form of art - the art of coping with fuzziness inherent in human knowing, the art of searching for meaning while stuck to apparently irrational life trajectories - trajectories that inevitably approach death and physical disintegration.

1.3.2 Awareness of Life as it Unfolds

The fuzziness in our understanding of society has roots in the self-referential nature of our awareness of human dynamics. It is an awareness of what happens inside and outside us as a *living movement in which we are also included*, without fixing it or standing apart from it. It is an awareness of life as it unfolds through each of us, through our society and through the universe, a profound awareness of human dynamics working within us. Such awareness implies an approach which is different from simply observing, fixing and comprehending social processes as if they are 'over there' (outside of the observers' mind). In becoming aware about the social dynamics, our experience and understanding of them remain inseparably connected with the innermost nature of each of us, and this experience and understanding gradually (or suddenly) transform one's individuality while taking hold of it.

Therefore, to grasp the fuzziness of our social experience, of what we know and understand about society, means to grasp the fuzziness of one's unique individual experience, of one's own knowledge about oneself. This kind of fuzziness changes (moves, evolves, transforms) together with the changes (movement, evolution and transformation) of each of us.

When fuzziness moves, transforms and evolves, we have a greater chance to see an increased number of its limitless embodiments and sharpen our awareness of:

- the dynamics of our inner nature
- the webs and whirlpools of our social interactions
- the evolving dynamics of the natural environment and our vital inseparability from these dynamics
- the ways in which life-supporting rhythm manifests through us and the environment
- the creative power of our inspiration and intuition
- the enigma of the spiritual continuity of existence.

Social fuzziology digs into the fuzziness of our understanding of all these phenomena as they cross our inner being, while responding to the turbulent dynamics of social life.

1.4 Socrates' Wisdom at the Origin of Fuzziology

I know that I know not (**Socrates**)

1.4.1 Meno's Paradox

In Plato's Meno Socrates explores what has been called Meno's paradox of learning:

If we don't know what X is, how can we recognize it?
If we can't recognize X, how can we learn what X is?

Socrates' discussion developed Meno's paradox into a more general paradox of human inquiry:

We know what X is. (Then we are not motivated to inquire into what X is.)
We don't know what X is. (Then we are motivated to inquire, but are frustrated by the paradox, since we cannot recognize instances of X, or what X is in general, to find out what X is.)

Discussions of Meno's paradox typically turn it into a proposition issued by a master (Plato or Socrates). In fact Plato's text describes an incident which dramatises processes of inquiry and proof. This social context is not mere decoration. It provides information for a social form of the paradox, which is crucial for social fuzziology.

The chief character debating with Socrates in the dialogue is Anytus, a historical character of great importance in Socrates' life. He was a leader of the 'democratic' party in the Athenian assembly which overthrew 'the Thirty', a tyrannical oligarchy who ruled Athens briefly in the chaotic years at the end of Athens' disastrous war with Sparta which was the death-knell of Athens as an independent state. He was also the leading force in the Assembly who moved and successfully argued against Socrates on the grounds of 'impiety', as a result of which Socrates was condemned to death. The context implies:

(1) Meno's paradox is not simply an academic argument, it was seen as a threat to the state. Socrates's 'proof' of his proposition took the form of an interrogation of a slave boy about geometry, in which Socrates' careful questions brought out mathematical knowledge from the boy he did not know he knew, 'out of himself'.

(2) Anytus, democrat and politician, did not like this argument, which was too democratic (Athenian democracy did not extend to slaves). He successfully accused Socrates of 'corrupting the young' by teaching them 'novelties'. In this case the 'young' was a slave, and the novelty he was taught was that mathematics is comprehensible to someone without formal education.

1.4.2 Socrates' Approach

Socrates' approach to Meno's paradox of learning included four steps: (1) generating hypotheses; (2) testing the hypotheses against examples; (3) philosophic examination of the hypotheses; (4) drawing out implications for learning and inquiring further.

Socrates' approach was a double-edged weapon. While examining what people (important men like Anytus) consider as known, Socrates draws out implications and asks questions that may lead them to contradiction (incoherence), circularity (presupposing what is at issue), infinite regress or other violations of epistemic norms. The same process addressed to the supposedly ignorant, like the slave boy, shows the opposite - that they may implicitly understand far more than they are credited with.

The trial of Socrates was full of paradoxes. Socrates, one of the most virtuous men who has ever lived, was convicted of 'impiety' and corrupting the young. His accuser was a 'sincere democrat' who found Socrates' freedom of thought and speech intolerable. Socrates contributed to the death penalty by refusing to ask for mercy or mitigation, claiming instead he should be rewarded by being given the freedom of the city, which was true but so angered his potential supporters that they swung to the other side. Nothing happened as it ought to have, according to the common reasoning, but exactly the opposite.

The chaotic situation produced two paradoxes: one of the most unjust legal decisions in recorded history, one of the greatest insights into the paradoxes of human knowing. The approach used by Socrates had made him aware of the famous paradox: *The less we know, the more certain and precise we are in our explanations; the more we know, the more we realise the limitations of being certain and precise.*

Although Socrates' wisdom was incomparably deeper and broader than the transitory knowledge of the majority of his contemporaries, he used to say with a proverbial humility: "*I know that I know not.*" The awareness that "*I know not*" allowed Socrates to reveal gaps in the 'precise' and 'certain' knowledge of his opponents. When the Athenians went to the famous Delphic Oracle to ask who was the wisest man in Athens, the answer of the Oracle was "Socrates". "But how can he be the wisest if he permanently tells us that he knows not," responded the crowd. "That is why he is the wisest among you," was the answer of the Oracle.

The acknowledgment of the fuzziness in human knowledge serves a stimulus for a lifelong inquiry and search for truth and wisdom; and it is this search that makes human life meaningful. Meno's paradox and the paradox of Socrates are at the conceptual basis of fuzziology.

1.4.3 Maieutic Inquiry

Maieutic inquiry (from the Greek word *maieutikos*, which means 'midwifery') is a method developed by Socrates; Socrates used to call himself a midwife who would bring about the birth of new ideas in people. The method implies asking people questions in such a way that it is possible to draw knowledge out of them - a knowledge that, according to Socrates, they already possess (Taylor, 2001).

Maieutic inquiry is based on the famous Socratic axioms that

- unexamined life is not worth living, and
- human knowing is limitless.

The practical realisation of maieutic inquiry occurs through a dialogue between two sides - one asking questions (the inquirer), and the other (the respondent) trying to answer them based on available pieces of knowledge. Both sides are interested in the process of inquiry: the respondent - to confirm the significance of the available knowledge, the inquirer - to reveal its limits and thus to facilitate emergence of new insights. If such an emergence occurs, the inquirer and the respondent move together beyond the limits of what was considered known by them before initiating the process of inquiry.

Socrates was convinced that one can always generate questions which push the boundaries of what is assumed to be known; so that these boundaries are never fixed. Every time the known is locked into patterns with rigid (non-fuzzy, crisp) boundaries, knowledge tends to become dogma, and the dogmas are no longer knowledge. The fuzziness of the boundaries demarking any domain of human knowing is a vital condition for its evolution and transformation. This was revealed by the wisdom of Socrates more than 2400 years ago. And not only this.

For Socrates the known appeared as a *symbiosis* (the Greek word for "co-existing") of many qualities, and the process of their understanding can be deepened with no limits. Such qualities always escape precise definitions, and therefore appear fuzzy to human mind. In maieutic inquiry these are 'individual' and 'social', 'subjective' and 'objective', 'internal' and 'external', 'concrete' and 'abstract', 'rational' and 'intuitive', 'partial' and 'holistic', 'local' and 'global', etc.

For Socrates (and for Pythagoras who lived 150 years before Socrates, it seemed clear that human beings must strive, at any cost, to understand the enigma of spiritual continuity of existence. What does this mean?

First of all, the great thinkers of the Ancient Greece believed that human life does not finish with death. People *can* be aware of life events, the experience of which proves that there are qualities in human nature that survive body's disintegration. If one cannot succeed in reaching such a degree of awareness, life appears meaningless - we come to life in order to die after a while, or create offspring destined to die also.

"*Life which implies death, how can it be called life*?" asked the ancient thinkers. *It is a hidden death rather then life*. It must have been hard for Socrates's wisdom to accept that nature could approve such a meaningless life for humans endowed with a gigantic capacity to explore and understand both themselves and the universe. It would have seemed reasonable for him to see humans as exponents of a much greater Life which extends beyond its physical manifestation. But it must have been clear for him also that human awareness of spiritual continuity of life does not come automatically. Its awakening needs efforts - genuine and persistent efforts on behalf of the whole triad of one's body, mind and soul supported by devotion and determination in order to reveal the immortal essence of Life before the moment of death of the physical body.

Fuzzy as it might appear from the standpoint of our science today, the exploration of spiritual continuity of life was at the core of Socrates' self-inquiry into this greatest enigma of human existence, and his famous "*I know that I know not*" reflects the soundless mystery of this enigma. Maybe it was Socrates' understanding of spiritual continuity of life that made him categorically reject the suggestion to ask for mercy when unjustly accused by the society in Athens - a society full of envy, hatred, suppression and ignorance. How far was (and continues to be) any society from understanding individuals whose thinking extends far beyond the average 'they-say' fuzziness of thinking of the majority of people and their rulers!

As a method for exploring the fuzziness of human knowledge which facilitates, through the skill of *maieuticos*, the 'birth' of new insights, the maieutic inquiry of Socrates is a useful research approach of social fuzziology.

1.4.3.1 Conditions Enhancing Maieutic Inquiry

Maieutic inquiry depends essentially on the active interaction of the inquirer and respondent. This kind of inquiry does not represent a problem-solving process; both sides do not search for ways to eliminate fuzziness from what they label as 'known' and 'unknown' about the issue(s) of their concern. *Fuzziness is an intrinsic characteristic of human knowing and cannot be eliminated from any stage of its evolving.* It is rather a kind of dialogue which helps the sides *dissolve* the impediments on the way to their understanding of the discussed issue(s) and thus loosen the knots into which they might have entangled themselves consciously or unconsciously by their prejudices, fixed ideas, borrowed solutions and delusions.

The interactions of sides involved in maieutic inquiry aim at liberating their creative potential from the pull of forces born out of human egocentricity and egotism, blind attachments and addictions, social brainwash or power-based manipulations - forces which are able to convert the fuzziness of knowing into hard-to-surpass ignorance.

The dialogue of maieutic inquiry unites rather than separates the dialoguing sides, and makes them act at-one when dealing with the limits of known.

Among the conditions that facilitate emergence of creative insights in an open maieutic dialogue are:

- *thirst* for understanding and knowing;
- *authenticity*, that is, freeing the mind from all kinds of biases, prejudices, dogmas and ultimate 'truths';
- *holistic questioning*, that is, asking questions that allow the process of knowing to go deeper and broader.

The first condition - *thirst for understanding and knowing* - relates to the proverbial ability of Socrates to inflame his students, to fire them with such a great passion to know that nothing seemed more important for them than the search for truth, a lifelong search undertaken together with their Master.

The second condition - *authenticity* - relates to Socrates' humble expression that 'the only thing I know is that I know not', and points again to the organic connection of maieutic inquiry with social fuzziology. Socrates' expression is not only an indicator of his humility and modesty. In the ability to prevent mind from formation of rigid patterns of knowledge, and thus to keep the process of knowing in a receptive and open state of changing its fuzziness, lies the secret of a wise person. Wisdom is authentic and the words of wisdom are not precise. But what they express does not appear fuzzy for those who can understand it. On the contrary, its message is illuminating; it can 'move' minds, hearts and souls of different people, evoke meaning in different situations and stimulate people's urge to know more about themselves and reality.

The third condition - *holistic questioning* helps not only to reveal the fuzziness in what is accepted as 'known for sure', but also takes the process of knowing deeper and further, preventing it from becoming crystallised into frozen patterns in mind. "*Never stop questioning*!" is the message of Socrates. *Answers live only for a short time, the questioning goes forever.*

1.4.3.2 Maieutic Way: From Knowledge to Wisdom

Some questions require yes-no answers ('Are you male?'), others are catechistic - questions to which the person asking it already knows the answer and is testing the other ('What is the name of the Greek philosopher who taught Plato?'). Such questions, which are common in most education systems, are a form of training in Aristotelian logic.

Often we know the answers of the questions we ask and yet continue to put them; these are questions which imply efforts - physical, mental, emotional or spiritual - in order to fulfill what their answers require. By asking such questions ever and ever again, we try either to hide our lack of will power for realization of what they urge us to do or to deceive ourselves that we are doing something while thinking about them.

Holistic questions - the questions asked by Socrates and social fuzziology - are entirely different. They are open, dynamic and inextricably linked ultimately to the fuzziness of the whole fabric of human experience and knowledge. They are questions endowed with power to transform the fuzziness of knowledge into the illuminating - enlightening, inspiring and soul-elevating - 'preciseness' of wisdom.

Maieutic inquiry of social fuzziology can be seen as an inquiry into conditions under which knowledge can be transformed into wisdom. There are crucial differences between knowledge and wisdom.

While wisdom needs the vibrant fuzziness of human thoughts, words and actions in order to inspire and evoke their creative understanding, knowledge constantly tries to reduce the fuzziness, substitute it with precise definitions or simply get rid of it.

Knowledge can be transferred, borrowed from books and experts, imparted and taught; wisdom is non-transferable, it is a unique individual treasure accumulated while riding on the tides and ebbs of life.

Knowledge is inevitably partial, it sets boundaries, hangs labels, separates and tries to generate precision - precision that always turns out to be meaningless when dealing with spontaneity of one's life unfolding. Wisdom is holistic; it accepts the unlimited - the timeless and the infinite - and sees clearly that the precise formulas and definitions never work in life.

Below are examples of some holistic maieutic questions:
Why do we exist on this planet?
What is the purpose of being born and then to die?
Is it possible to escape the death sentence that each of us was born with? How?
What is the meaning of one's life?
Where do our thoughts and feelings come from, our emotions and longings, aspirations and dreams?
What propels the life-sustaining rhythm of each person's heart?
What makes the cells and organs in the body not stray away but function in accordance?
From where come the waves of inspiration?
How do we intuit?
What does enlightenment mean?

The above questions directly zoom into the fathomless depth of our essence as creatures endowed with the potential to comprehend reality.

Attempts to answer a maieutic question may lead to other questions, and their answers may fire another inquiry. The maieutic way of exploring fuzziness of human knowledge - its nature, sources and dynamics - never ends. Nor is there an end to the emergence of new insights about the truth of existence to those who, like Socrates, see the mission of their lives in revealing it.

While using the technique of maieutic inquiry, social fuzziology continues to explore the same process to which Socrates devoted his life - the process of transforming fuzziness imbedded in human knowledge into insights that spring out of human wisdom.

1.5 Principles of Fuzziology: A Social Perspective

1.5.1 Principle of Incompatibility

As the complexity of a system increases, human ability to make precise and relevant (meaningful) statements about its behaviour diminishes until a threshold is reached beyond which the precision and the relevance become mutually exclusive characteristics (Zadeh, 1973). It is then that fuzzy statements are the only bearers of meaning and relevance.

This principle was used by Zadeh for extending the applicability of his fuzzy sets theory and fuzzy logic to the analysis of complex systems. Given the high degree of complexity of all known human societies, this principle applied to social fuzziology states:

The more precise the language that sustains and describes human societies, the less socially effective (able to maintain the integrity of social relations) and the more distorting (unable to describe adequately the meaning of social forms, processes and actions). The more complex and dynamic the context of explanation, the more essential will fuzziness appear to the human understanding which constitutes and explains it.

1.5.2 Principle of Connectivity-in-Dynamics

No thing and no being can exist in itself or for itself but only in a dynamic relationship with other things and beings.

This principle relates to the integrity of existence vitally supported by universal dynamics, whose creative, sustaining or destructive powers are constantly demonstrated at different scales of the manifested world. It is through these dynamics that everything that exists, from an elementary particle to a gigantic galaxy, becomes connected in an all-embracing web of relationships.

The application of this principle to society in social fuzziology suggests that:

Social meanings, social relations, social identities and social actions exist irreducibly in open, dynamic networks of relations.

1.5.3 Principle of Fractality

The geometry of nature is fractal and reveals itself as self-similar structures at different scales of manifestation.

This principle is at the basis of Mandelbrot's theory (Mandelbrot, 1982) of fractals and demonstrates the way self-organisation works while unfolding the complex dynamics of nature. Self-similarity is a kind of fuzzy repetition; each scale has common features with every other, and yet there are noticeable differences.

Fractals are inherent in the holistic unfolding of individual, social and existential dynamics: the macrocosm is a projection of the microcosm; the external world of individuals is a projection of the inner world of their experience, which fractally repeats that outer world; each level of development of consciousness has similarity both with the previous (less developed) and the next (more advanced) levels and yet has its own distinguishable characteristics, its own strength and weakness.

The principle of fractality applies directly to social phenomena:

'Society' has equivalent significance and complexity at every level, from the biota over billions of years to individual nations, groups, classes, down to 'individuals'.

From a 'fractal' point of view an ‘individual’ therefore is always a social form, intrinsically connected to social forms and patterns at higher levels, containing social forms and patterns within.

Ethical decisions at local levels - eg project-based, work-place based decisions - will be similar to ethical issues at higher levels, up to the scale of planet, involving humanity and all other species.

1.6 Impossibility Theorems

1.6.1 First Impossibility Theorem

It is impossible to eliminate fuzziness from any explanation that tends to make sense of

- *the wholeness of the existential dynamics*
- *the infinity of their manifested activities.*
- *the immensity of their potentiality to create.*

The validity of this statement follows from the first two principles of social fuzziology. According to the Principle of Connectivity, the wholeness of existence, its manifested activities and its creative potential are results of an all-embracing connectivity of everything that exists, that moves, changes and transforms in a gigantic self-organised Web of Interdependent Dynamics. According to the Principle of Incompatibility, it is impossible to offer precise and yet meaningful explanations related to the overwhelming complexity of this web. Hence, any possible explanation that makes sense of the integrity of existential dynamics, their unlimited actual or virtual appearance (as "manifested activities" or "potentiality to create") inevitably contains fuzziness.

The First Impossibility Theorem prevents fuzziology from looking for and from designing techniques to eliminate the fuzziness of our knowledge of social complexity; such techniques cannot be found. The fuzziness of social complexity has its deep roots in the very essence of existence - an essence whose self-propelled unfolding makes the universe "incomprehensible" to our "frail and feeble minds" - expressions used by Einstein when describing his religion.

"My religion consists of a humble admiration of the illimitable superior spirit who reveals himself in the slight details we are able to perceive with our frail and feeble minds. That deeply emotional conviction of the presence of a superior reasoning power, which is revealed in the incomprehensible Universe, forms my idea of God" (quoted in his New York Times obituary of April 19, 1955).

Human "frail and feeble minds" are products of the unfolding of the mysterious essence of existence. Therefore, their fuzziness is not something 'over there' that can be objectified, rationally defined and then studied and modelled; it is deeply inside each of us and, therefore, escapes the grasp of our reasoning. It can be felt, experienced and eventually realised in life. Being out of the realm of logical formulations (no matter what kind of logic we decide to use, be it inductive, deductive, abductive, binary, multi-valued or fuzzy), the journey into the existential mystery needs 'preparation', in which the reasoning power of the human mind plays an important role, to co-ordinate sense impressions, perceptions, sensations, feelings and emotions into a meaningful whole.

Fuzziology acknowledges the irreducible fuzziness of human knowledge about the essence of the existential dynamics. The awareness of this fuzziness activates the potential of fuzziology in construing reality where the conscious revelation of our deep, inner experience plays the paramount role in making sense of existence, not the intellectual speculations about the outward, 'objective' world as perceived through our senses. The information from our senses inevitably passes through mental and emotional filters, consciously or unconsciously established in the process of socially informed interactions. Some of these filters can irreversibly distort the sensory information to such a degree (as the result of bias and prejudice, brainwashing or propaganda, attachments and delusions) that it entirely ceases to help people navigate the social complexity of their lives.

All of these considerations apply equally to the irreducible fuzziness of human sociality. This suggests one particularly important consequence for social fuzziology in its relations to the current set of disciplines which purport to describe and explain social life:

The disciplines of the social sciences as subdivided from each other and separated from the totality of the field of human knowing are inherently incapable of understanding the essence of social life, or activating creative processes of interaction. The main models, terms and categories of the social sciences, where they aspire to precision and crispness, will inevitably provide a distorted and diminished understanding of social life. They need to be 'fuzzified', that is, enriched with rooted-in-life approaches aimed not so much to explain social reality in scientific terms but rather activate creative insights for understanding and dealing with its paradoxes and enigmas.

1.6.2 Second Impossibility Theorem

It is impossible to understand and deal with fuzziness related to a higher (more developed, expanded) level of consciousness from the point of view of a less developed level.

The validity of this statement follows from the Principle of Fractality when applied to the unfolding of existential dynamics. From their manifestation at the scale of non-animated nature (described by the ancient thinkers as being built by fire, light, air, water and earth) dynamics unfold to express themselves at various scales (levels) of animated nature, at the scales of plants, animals and humans. The unfolding of these dynamics runs parallel with a self-propelled expansion and growth of complexity at each scale of manifestation. There is a stunning diversity at the level of minerals, and also at the levels of plants and animals.

The complexity at each level of existential dynamics' unfolding cannot be reduced to the complexity of the previous level: animals' lives are of a higher order of complexity than the life of the plants, which are themselves much richer and diverse than the 'life' of minerals. When dynamics enter the human scale, it is human consciousness (as a holistic experience and awareness and knowing of our own nature and the nature of reality in which we exist and evolve) that expands and grows.

The fuzziness of knowing at each level of development of human consciousness can hardly be grasped from a lower level of consciousness. What may appear as a 'fuzzy mess' for individuals with a certain level of development of their consciousness can be seen as saturated with meaning, if they exert effort and succeed in developing a higher level of awareness and intelligence, or/and in sharpening their capacity to think, feel and experience holistically, rather than solely from a narrowly established point of view ('worldview').

As the example of Plato's Meno illustrates, a mere accumulation of knowledge or degrees in education does not mean wisdom or higher levels of consciousness. The 'well educated' Anytus believed he was superior to Socrates, and far above the level of the slave boy. Yet as Socrates demonstrated, the slave boy's eager and open mind could discover a mathematical proof, when properly led by the wisdom of Socrates, while Anytus's arrogance meant he was unable to understand either the slave boy or Socrates.

Far from social elitism being an illustration of this theorem, it serves to reinforce the distinction between knowledge as an accumulation of facts, theoretical explanations and practical skill and wisdom as holistic insights into existence born while living and experiencing existential dynamics in their all-embracing integrity and infinity.

What the wisdom of Socrates could grasp was far beyond the understanding of many of his contemporaries. And the enigmas of life which appeared fuzzy to Socrates and kept the passion of his inquiry alive till the day he was unjustly accused and killed, quite possibly never bothered most of the Athenians at that time.

The fuzziness inherent in the deepest spiritual wisdom of the ancient Vedas, considered the oldest written text on our planet (coming to us in written form between 4000 to 6000 years ago) is almost impossible to grasp with the level of consciousness of our generations - consciousness deeply immersed in a constant pursuit of materialistic acquisitions, selfish accomplishments and anti-human manifestations of ego-centred power.

One message of the Second Impossibility Theorem is that the life-threatening fuzziness of all the serious ecological, economical and social problems, which today's humanity creates, and by which more and more people are tormented, can hardly be solved using the present level of consciousness typical for the rulers of the developed capitalist 'democracies', driven as they are by an insatiable thirst for money and pleasures, competitive advantages and power.

1.7 Social Nature of Fuzziness

1.7.1 The Double Face of Fuzziness

In writing this book we were strongly drawn to social fuzziology for ethical motives, by a desire to help the world to be a better place. We believe the current tendency to separate science and ethics has proven a dangerous mistake, an experiment which has sufficiently proven that the planet cannot survive a science and technology, a politics and economics without an ethical and spiritual dimension. In this context, recognition and use of fuzziness is an ethical act, an ethical imperative.

Yet it is also the case that fuzziness is not always good or bad, nor is non-fuzziness always bad or good. Ethics is not only a matter of mind, but involves other strengths of human nature. Qualities of mind cannot provide the sole criteria for judgements of good and evil. At the same time, ethical systems themselves are products of human minds and judgements. They can be more or less fuzzy, more or less adequate to the complexity of other kinds of understanding of inner and outer experience. But human values like beauty, justice and love come from other sources as well. Fuzziness must be judged by ethical values outside itself, even though we may use our experience of fuzziness in our continual attempts to refine these values.

There are three principles describing the variable ethical value of fuzziness:

- there is no absolute fuzziness: it exists only in contexts and relationships to human experience and knowledge, thinking and speaking, feeling and acting, etc. at a particular moment;
- fuzziness is always in a dynamic state, and finds value in its direction and momentum not just in its particular position;
- fuzzily expressed contexts, relations and tendencies have ethical value that grows out of the totality of life.

Fuzziness is a treasure, and the fuzziness of our thinking is inseparable from our joy in creativity. Attempts to crush this fuzziness in individuals or in whole populations are tantamount to killing life itself, and killing life is surely wrong. Yet it is also the case that to attempt to understand a situation or problem that is experienced as too complex by generating less fuzzy solutions can be immensely creative. The opposite response, to simply accept an amorphous fuzzy understanding, can be passive and lacking insight.

We can capture this doubleness in the form of two contrary propositions:

- creativity and insight can be powerfully released by trusting fuzziness of mind;
- creativity and understanding can be unleashed by an energetic journey of understanding from greater to less fuzziness, especially through journeys that go forwards, into transcendence of fuzziness, not backwards, to its denial.

To illustrate the complementarity of these two propositions in relation to fuzzy logic as developed in engineering contexts: applying principles of fuzziology to fuzzy logic can restore its creativity and explanatory potential, while applying fuzzy logic to social complexity without renouncing that complexity could be stimulating and productive.

1.7.2 Fuzziness and Power

Human social relations are characterised by the effects of power - acts of domination and resistance in complex forms. Social fuzziology needs to be always alert to effects on ethics and truth of interrelations between power and fuzziness. In general there is an inverse relation between power and the value of fuzziness: where social relations are strongly determined by power, fuzziness becomes *delusive fuzziness*, and non-fuzziness becomes *dogmatism.*

There are many examples of this proposition. For instance in the first year of his Presidency, George Bush promoted two proposals, one which seemed to welcome fuzziness too much, the other to respect it too little. He opposed the Kyoto Protocol to control greenhouse gas emissions, then close to finalisation, on the grounds that there were some 'doubts' about the exact extent of global warming and the contribution of human-produced greenhouse gases.

The Bush administration proposed to wait for further research to resolve these doubts. In response to this position, a group of science academy's issued a consensus statement urging the Protocol be adopted. They noted the existence of uncertainty ('There will always be some uncertainty surrounding the prediction of changes in such a complex system as the world's climate') but after careful deliberation they accepted a figure of 'at least 90% certain' that temperatures will rise between a predicted range (Editorial in *Science*, 2001).

Bush reacted in the opposite direction to criticism of his proposal for an anti-ballistic system to replace the previous Arms Treaty. Faced with criticisms that the technology did not yet exist to make such a scheme feasible, he responded with faith that it was possible, and that 'buckets of money' should be devoted to this dream.

In the first case, the relative fuzziness and uncertainty of the arguments in favour of the Kyoto Protocol became his reason to ignore them completely, in spite of the consensus of the scientific community that the fuzziness and uncertainty existed within manageable limits which were understood and accepted. In the second case, the fuzziness and uncertainty of the arguments in favour of his anti-ballistic system became a reason to back the creativity of his scientists to tackle and solve the problems.

The Bush administration used more than fuzziness. In July 2001, the Pentagon publicised a 'test' of the feasibility of the available missile technology, in which two out of four missiles struck their ballistic target. 'They hit a bullet with a bullet, and it does work!' proclaimed Republican Senator Trott. But then it was revealed that the target missiles all carried beacons to allow them to be tracked from space, prior to the radar system cutting in. Pentagon officials admitted the beacons were 'a big help', and that enemy missiles would not have come equipped with them. (Riley, 2001).

If this report had been prepared by a scientific institute it would be declared a fraud. It is no surprise to find such things occur with the amounts of money and vested interests at stake. The public has grown cynical about the cynicism of politicians. But the case makes a less obvious point about fuzziness and its interactions with power. The technology Bush promoted on relies on precision, and therefore produces either its opposite - uncertainty, or fraud. Fraud in a double guise - manipulative fuzziness about what technology might do, concealing the actual conditions of the test - is a consequence of the interaction of fuzziness and power. Non-experts do not know the exact details of what fraud was perpetrated, but the general public know this 'principle' of social fuzziology: *when power and fuzziness are related, truth is the first casualty.*

1.8 Human-centred Sources of Fuzziness

1.8.1 Mind as a Source of Fuzziness

The greatest source of fuzziness is our mind. This powerful rational thinker never ceases to divide the whole of reality into fragments in order to analyse, classify and label them, and then to toss or scrap together to piece out a world, which has very little to do with the unbreakable wholeness of reality, a wholeness we inseparably belong to.

The world we piece together from fragments made out of perceptions - sensations and thoughts, serves to provide partial and therefore distorted models of reality. These models represent a world as-perceived, a human-made world, and not a world whose natural evolution has brought us to existence and with which we are linked through an umbilical chord of vital and impossible-to-separate connections. All our models deal with parts of something that we perceive as over-there, as surroundings considered to be used for what our ego-centred minds label as meaningful.

An army of scientists, engineers, economists, politicians and philosophers are involved in adapting many distorted models for predicting and exercising power over the unfolding dynamics of reality. Although we know that complex dynamics of reality are beyond our ability to predict and control, we 'do our best' to mutilate reality so that it could be pushed into Procrustean beds of reductionist models. The applications of these kinds of models have made both nature and society vulnerable; this is clearly demonstrated by today's ecological disasters and continual worsening of socio-economical conditions for the largest and ever-increasing part of society.

The rational mind can never move beyond duality. It either selects something while rejecting its opposite (as in black-and-white thinking when using binary logic) or accepts both the opposites up to some degree (as in fuzzy thinking when using fuzzy or probabilistic reasoning). The dualistic nature of the rational mind is so strong that it is unable to transcend it alone. The best it can do is to reconcile the opposites by eliminating one, or blurring them both.

Following the black-and-white approach in thinking ('*either* A *or* not-A'), we can be easily entrapped in routines, stereotypes, prejudices and habits that become a source of fuzziness which eventually makes us incapable of authentic experience. All our 'understanding' is constantly filtered through already established mental patterns.

Following the fuzzy logic-based approach in thinking ('*both* A *and* not-A' up to some degree), we may agree too much to everything the others say, and this can push us towards compliance and indecisiveness. When everybody is right, the uncritical acceptance of the fuzziness accompanying other people's thoughts makes it hard for us to generate our own creative ideas. It is the polarity of opposites, contradictions and clashes of opinions, that provides human mind with dynamics (forces and energies) necessary for transcending the opposites. These dynamics manifest in mind as an urge to search beyond the plane where the opposites clash. Without such an urge, the mind can be entrapped into stasis, stuck in repetition or mesmerised by illusory thoughts and dreams.

"*Do not reject anything! But do not remain with anything either! Go beyond*!" is written in the Vedic scriptures. In our context these words of wisdom say: "When searching for understanding, be ready to go beyond logical rules and restrictions, no matter how soft (fuzzy, probabilistic) or hard (binary, deterministic) they are!"

1.8.2 Ever-emerging Desires

Another powerful source of fuzziness is rooted in our ever-emerging desires of various kind - from simple physical desires which we share with other animals to much more complicated desires specific for human nature. Every desire agitates the mind and distracts the process of concentration indispensable for an act of understanding to be productive. The stronger an emergent desire, the higher the degree of agitation it stirs up, the less the degree of concentration of mind; the less the degree of concentration, the fuzzier the process of thinking, the lower the degree of understanding.

Most of the desires self-propel their intensity - the more we try to satisfy them, the higher become their demands. The way of moderation, the 'middle way' in terms of the Buddhist thinkers, is hard to follow when the fire of desires is burning inside us and making our minds restless, turbulent, obstinate.

The restlessness and turbulence of minds are permanently intensified by the stress in which we live due to the competitiveness inherent in today's society and the helplessness of the majority of us to get out of the social boxes and cages, in which we have been pushed by economic forces too strong to withstand.

Although the strength of passion with which we pursue truth and understanding is a powerful stimulator and energizer of thinking, understanding also needs 'peace of mind', a mind which is calm and cool, composed and collected.

Paradoxically, while being sources of fuzziness, mind and desires are, at the same time, key factors for overcoming it, especially if the fuzziness relates to problems deeply rooted in human experience.

1.8.3 'Golden Rules' of Fuzziology

However concrete and precise human actions might appear, in the context of social complexity their effects are 'fuzzified' through a multitude of consequences, both known and unknown, open or hidden, erupting immediately or after a hard-to-predict interval of time. Therefore, almost every action within social complexity is fuzzy. We can hardly be certain about the consequences they lead to.

The way to cope with the fuzziness of understanding consequences of social actions is to be aware of it, not pretend that it does not exist or hurry to substitute it with straightforward cause-and-effect explanations. Our haste to offer such explanations when justifying the application of any economic, technological or political decision is responsible for the most serious maladies of today's society: environmental destruction, disconnection of economy from society, extreme inequality in the distribution of wealth, degradation of work, etc. Awareness of fuzziness when dealing (working, acting) with social complexity goes hand by hand with our drive to understand the nature and dynamics of every incarnation of this complexity as it appears in our lives, both individual and social.

In the context of the inseparability of human understanding and social complexity, an inseparability which is at the very core of social fuzziology, one can visualised the subject of social fuzziology through the wings of the famous butterfly attractor of Lorenz - one of the creators of chaos theory. From the wing of understanding complexity we move towards the wing of working with this complexity, and from there again to understanding, and then again to working, and so on, in a never-ending attempt to realize the uniqueness and infinity of our potential to think and act.

In order to keep Lorenz butterfly moving, so that each flap of its wings might be able to bring forth not only a "hurricane" or "tornado" (metaphors used by Lorenz when describing the butterfly effect in chaos theory) but also real fruits of human creativity, social fuzziology offers three 'golden rules', not to follow blindly but to consider consciously as practical tools for strengthening individual awareness:

- Avoid neglecting the unknown, denying it, turning away from it or trying to make ourselves and others believe that it is really known in order to organize, dichotomize and impose rules on it. The unknown manifests through spontaneity of any novel expression of human creativity

- Avoid clinging to a need for certainty, definiteness and order or to ideas and practices that are familiar, commonsensical or accepted as true by an assumed majority. Remaining attached to what is certain and familiar suppresses idiosyncrasy of the human potential for self-realization.

- Avoid fighting with the complexity of life dynamics, no matter where they manifest - in our inner nature or in the world around. The way to avoid being a slave or a victim of these dynamics is through a constant pursuit of understanding how they work and how to apply their infinite energy for growth in intelligence and spirit.

1.8.4 Danger of Dogmatising Fuzziness

Fuzziology warns us against the danger of dogmatising the fuzziness inherent in human understanding. When making dogmas out of fuzziness, we can easily kill its capacity to move - to change, evolve and transform together with the changes, evolution and transformation of the process of one's knowing; when we keep its dynamics alive, we never stop strengthening our capacity to learn and know.

There is almost no danger of following strictly a precisely described instruction - manual, algorithm - that helps one run a dynamic technological process (or a sophisticated engineering system). Enormous danger is involved in choosing or being forced to choose to follow dogmatic interpretations related to somebody else's understanding of life complexity. The manuals and algorithms helping to run an engineering system deal with the limited sphere of an artificial (human-made) reality, the logic of the statements about which can be tested and verified in an objective way. The fuzzy knowledge necessary for dealing with complexity of one's every day existence, while living in this complexity, inevitably covers a much larger space of possible interactions and interpretations. The truth of any piece of fuzzy knowledge about the living reality can be tested and verified subjectively, while experiencing individually and socially the immediacy of its limitless variety of dynamic manifestations, without pushing this variety into precise manuals and algorithm, formulas and equations, rigid rules and dogmas.

As far as each individual life unfolds in a unique unpredictable way, any voluntary or externally imposed choice to live following precise recipes impedes the possible emergence (discovery, creation) of a multitude of other (different) ways of dealing with the life complexity. The suppression of human creativity causes misery and anguish, blocks the natural flows of individual and social energy (be it physical, emotional, mental, spiritual), removes all the joy and gladness from human life and eventually destroys it.

It is impossible for an individual to be other than oneself - to live with one's own fuzziness of understanding and struggle with it every time when trying to grasp its bottomless sources. One suffers when fuzziness leads to delusion and pain, and feels happy when it triggers new insights and inspiration. Who knows, maybe this is the most efficient way of learning how to understand and cope with the enigmatic fuzziness inherent in the meaning and mission of one's life.

The proverbial expression of Nietzsche "*Follow not me, but you*" has a deep existential meaning. One cannot be transplanted into the inner space of another person. If we follow others, instead of being ourselves, we lose the creative sparks born out of our own struggles to understand the fuzziness of ourselves and make the light of our own unique individuality shine. Without this light, there is no self-awareness, no wisdom.

It is a grievous mistake to imitate someone else's fuzziness in experiencing the events of life. This mistake may result in fatal conflicts between the inner nature and mind of an individual (confusion in thinking), the inner nature and heart (confusion in feelings), and the inner nature and soul (confusion in experience and search for meaning and light in life). When imitating others' ways of living, thinking and acting, we are losing the most precious gift we have - the freedom to be ourselves, to express ourselves and realize our potential.

1.8.5 The 'Drowning-Man' Paradox

Any act of understanding is an act of emergence of meaning(s) out of the fuzzy thoughts constantly 'swarming' in one's mind. In the moment when a meaning emerges, we try to use it in making sense of what we experience - what we see and hear, touch and smell, read and contemplate, create and discover. Of course, this meaning is fuzzy, as it has emerged out of the fuzziness of our interactive thoughts. Once emerged, the fuzzy meaning immediately reflects the way we think. And as far as thinking is always coloured with feelings, the fuzzy meaning affects our emotional life too.

By influencing the ways we think and feel, the fuzzy meanings we create self-propel their growth - they act as magnets attracting more and more thoughts and emotions, which support them, and thus make them more definite, more categorical, more stable. For example, after Mandelbrot presented his idea of fractals, those who work in fields of research different than mathematics or computer science constructed their fuzzy meanings of fractals and actively started to generate ideas and accumulate information confirming these meanings, so the latter gradually became more definite and stable.

In the overall fuzziness of human understanding of social complexity, the tendency of any emergent fuzzy meaning towards stability reflects humans' own gravitation to stability, to something that is considered familiar, secure, habitual and known. In its extreme, this tendency leads to a paradox, which we call the drowning-man paradox:

The fuzzier one's understanding, the stronger one's attachment to what seems non-fuzzy in it.

Social life shows many instances of this paradox. For example, the denser the fuzziness of understanding dynamic complexity of a social situation by some governing body (managers, leaders, governors), the higher its willingness to implement stringent, that is, non-fuzzy methods of control. Unfortunately, the use of these methods only aggravates the existing problems and makes their fuzziness incomprehensible.

If the increase of economic and social turbulence in a state is beyond a government's ability to understand its nature and sources, the government sticks to bureaucratic methods of control. When an organisation is in a process or restructuring, the bureaucrats in it become extremely active: the number of rules, restrictions and regulations they produce dramatically increases. Although the rules, regulations and control actions appear crisp - black-and-white, categorical, rigid - they often lead to vicious circles, blockages on the innovative ways of thinking resulting in a further increase of the overall fuzziness in understanding and coping with the organizational complexity.

Before the collapse of the soviet model, Andropov's regime desperately tried to put into practice non-fuzzy (KGB-like) methods of dealing with socio-economic complexity, and this turned the life of the ordinary people into a hell.

Routines, prejudices and biases in human thinking also illustrate the spread of the 'drowning-man' paradox in society. We cling to rigid patterns of thinking because we lack will and courage to openly explore the sources emitting fuzziness in our own understanding of life. Every time we cling to the things others preach and teach, taking the fuzziness of their understanding as a 'precise' recipe and ignoring our own lessons in understanding complexity, we are captives of the 'drowning-man' paradox.

To avoid this paradox does not mean to use only unstable, easily changeable, superficial and open-to-manipulations meanings; these kinds of meanings can hardly trigger any earnest process of contemplation. In order to 'move' our understanding towards a deeper and broader grasp of social complexity, the emergent meanings need to be beyond duality of the labels 'stable' or 'unstable'; they can be used for generating hypotheses, concepts and emotional attitudes, without hardening or transforming these concepts and attitudes into dogmas. In other words, meanings need to freely emerge and dissolve in response to the changes occurring in reality of our experience.

1.9 Transcending Duality and Non-duality

When studying fuzziness of human understanding, social fuzziology looks for ways of transcending both duality and non-duality inherent in thinking.

With duality the mind constantly asserts

A is true *and* (therefore) not-A is not true, *or*

A is not true *and* (therefore) not-A is true.

Fuzzy logic based non-dualism accepts that

both A *and* not-A are true (up to some degree).

Spiritually enhanced non-dualism rejects the existence of dualism itself, maintaining that reality is an illusion (samsara in Buddhist terminology):

neither A *nor* not-A are true

Both duality and non-duality keeps the mind entrapped in logical speculations, and may lead to confrontation, one-sidedness and error (in the case of dualistic thinking) or to compliance, confusion and passivity (in the case of non-dualistic thinking).

Any insightful act of understanding vitally needs the energy of polarities expressed in dualistic thinking, as well as serenity accompanying their reconciliation in the framework of non-dualistic thinking. The way to take advantage of dualistic and non-dualistic thinking simultaneously is through expanding our consciousness so that we can flexibly switch from one to another mode of thinking, without being attached to either.

1.9.1 Polarisation

Paradoxically, it is common to find the greatest polarisation in situations where the two parties seem to have much in common, and strong motives for reconciliation. In these situations, which are as common and dangerous in differences between lovers, partners or dear friends as between enemies or warring groups, the signature is the intense passion and pain that seems to accompany key moments of the discussion, as though disagreement and difference are not intellectual matters but acts of betrayal. The two parties tend to occupy extreme positions, seemingly in antagonism to the other, often expressed in extreme language.

Superficially this may seem to be a case of the danger of non-fuzzy positions, expressed in non-fuzzy language. But on closer inspection the respective positions reveal themselves as not precise, only extreme. In fact they are normally full of internal contradictions, and 'wildly imprecise' in their extremism. Polarisation is a kind of bifurcation, not precision. Fanaticism is an extreme manifestation of this kind of dense fuzziness in thinking, when human ability to move beyond an established dogma is entirely blocked.

One indicator of this state is that the points of difference do not remain fixed, able to be argued against one by one in search of a compromise or resolution. If one point is agreed on, it quickly ceases to matter, as each participant finds another reason to fight the other to the death. The principle of polarisation is highly creative to a destructive end, producing a constant stream of reasons for the struggle to continue, but never to find a common way out of it.

Another indicator that this polarisation is very different from ordinary non-fuzziness can be seen in the fact that sincere attempts by outsiders or mediators to be conciliatory and fuzzy, to look for common ground between the two positions, to seek a consensus, will often be greeted with equal anger, generating a new position equally opposed. So 'fuzziness of thinking' as conventionally understood is not a road to consensus, at least in conditions of intensity and polarisation.

In states of relationships where the emotional intensity is high and issues are confused, ('far from equilibrium' states of relationships, feelings) then bifurcations may occur, taking the form of cascades of oppositions and polarisations. These polarisations are not logical structures (such that from a basic difference in premises, many other differences are logically entailed) but generative, able to produce new elements of difference out of any random phenomenon, still charged with much of the intensity of the original difference and able to act in its stead.

When the process of polarisation reaches this stage it is no longer possible to identify a premise that can be made fuzzy and brought into consensus (eg 'some Jews are good people' for an Arab, 'some Arabs love their children' for a Jew), because the differences are like the Hydra's heads, which Hercules tried to cut off, but when he cut one off, two more grew in its place.

In such a situation, mediators (someone trying to achieve a 'fuzzy consensus') will fail, and may be at risk to their life as well, as has been the fate of many people in Northern Ireland who tried to cross the boundary between Catholic and Protestant. Socrates was another such victim.

1.9.2 'Transcendent' Logic of Social Fuzziology

The situations described above are so common, and so dangerous that social fuzziology needs to theorise them in a way that is true to them, and leads somehow to good outcomes (greater health, greater beauty, greater love, greater wisdom) if this is possible. The ideas of social fuzziology must be useful: able to help a world that constantly collapses into bloody wars of brother against brother, oppressor against oppressed. What kind of thinking, what kind of logic can help?

Part of the answer may be found in a kind of logic we call *transcendent* - a logic where even the duality between fuzziness (*both* A *and* not-A are true up to some degree) and its negation - the lack of fuzziness (*either* A *or* not-A is true) ceases to exist, and researchers find themselves in a state of creative nescience or creative emptiness, characterised by extreme openness and responsiveness. In such a state we are ready to experience new dimensions of reality or discover new possibilities and meanings.

This form of logic is similar to what Kauffman calls 'virtual' logic: "Virtual logic is not logic, nor is it the actual subject matter of the mathematics, physics or cybernetics in which it may appear to be embedded... It is the pivot that allows us to move from one world of ideas to another" (Kauffman, 1997).

The Vedic maxim "Do not reject anything! But do not remain with anything either! Go beyond!" encapsulates virtuality of the transcendent logic and points to its possible use in social fuzziology as a mind-energiser, as a stimulator of human creativity and catalyst for seeking mutual understanding and social harmony.

In the endeavour of fuzziology to transcend the duality between fuzziness and non-fuzziness lies an essential difference between fuzziology and pyrrhonism - a radical skepticism initiated by the ancient philosopher Pyrrho of Elos (4th century

BC). Pyrrhonism postulates that certainty of knowledge is unattainable, so there are no ways to go beyond its inherent fuzziness. For fuzziology such ways exist and social fuzziology aims at exploring them in the context of individual and social realisations of human dynamics.

1.10 Consciousness Resonance

1.10.1 The Possibility Theorem

We can understand as much of the world as we have developed and realised within ourselves.

The validity of this proposition follows from the Principle of Fractality and from the Second Impossibility Theorem. The Principle of Fractality allows us to understand why the macrocosm mirrors the microcosm and the world outside reflects the world inside us, through a self-similarity that is never identity. The inner world is made not only of our senses, feelings and thoughts shaped into images, ideas, aspirations, expectations, hopes, dreams, but also of our deep spiritual attitudes and beliefs. We perceive the world around us through all of them. The power of our will is also in the inner world, together with our infinite potential to create and realise ourselves in innumerable activities. We never cease to modify the external world through actions emerging from our inner worlds.

The external world also affects the world inside us. The lower the level of consciousness, the stronger the influence of the external world, the more silent the voice of the inner world and the weaker our spiritual drives for self-realisation.

From the Second Impossibility Theorem it follows that when we grow in consciousness, we are able to see more of its projections onto the world around us, to develop and realise outwardly more of our inner potential to create. It is then that another type of fuzziness, inaccessible from the previous levels of consciousness, starts to irritate and challenge our minds and souls.

1.10.2 Journey Beyond Limitations of Fuzziness

In one of his book B. Greene, a professor of physics and mathematics at Columbia University, writes: "Although we are technologically bound to the earth and its immediate neighbors in the solar system, through the power of thought and experiment we have probed the far reaches of both inner and outer space. During the last hundred years in particular, the collective effort of numerous physicists has revealed some of nature's best-kept secrets. And once revealed, these explanatory gems have opened vistas on a world we thought we knew, but whose splendor we had not even come close to imagining." (Greene, 2000)

As far as the process of revealing "nature's best secrets" never stops, what "we think we knew" yesterday inevitably changes today, and new vistas "whose splendor we had not even come close to imagining" constantly open to those who are thirsty for knowing.

The fuzziness of knowing never ceases to exist. And this is a paramount characteristic of human knowing, which challenges humanity and constantly propels its search for truth and understanding the secrets of reality.

Fuzziness has its own dynamics and potential to 'pull itself by its own bootstraps". When our understanding deepens, the dynamics of the fuzziness tend to shrink; when our consciousness expands and our minds 'jump' into a higher level of reasoning, the fuzziness pulls itself also at the new level. So, fuzziness always accompanies the process of our understanding.

While exploring fuzziness, fuzziology reveals also ways of transcending it and thus expanding the field of the human inquiry. The fuzziness of understanding problems emerging out of complexity of life as it unfolds cannot be resolved at the same level of knowledge that we have when these problems appear. (One can see here an analogy with the Gödelian problems in mathematics and other fields of knowing - they cannot be resolved using the same assumptions under which they have appeared). Only when our consciousness is expanded or 'raised to a higher level' of reasoning and understanding, does the tension fades and the problems, being seen in a new light, cease to be problems. When problems dissolve, we say that the fuzziness related to them has been transcended.

1.10.3 Activating Consciousness Resonance

The qualitative jump in consciousness to a higher level results in transcending fuzziness accumulated in one's experience and knowing related to lower levels of consciousness. As far as consciousness is a holistic characteristic of human dynamics and not only a product of mind, the growth of consciousness is possible when the factors responsible for the integrity of all three inseparable constituents of human individuality - body, mind and soul, become simultaneously activated. We shall refer to this simultaneous activation as a *consciousness resonance.*

Consciousness resonance involves all factors responsible for human integrity. What are these factors? First of all, factors which contributes toward keeping the human body healthy and human mind capable of thinking and deciding, regardless of the preferred logic, be it fuzzy, binary, inductive, deductive, abductive, etc. But these factors are not enough! Consciousness resonance cannot occur when neglecting the *soul factors*: among them sensitivity and responsiveness, awareness and ability to stay awake, passionate desire to get out of the 'attractor' of egocentric thoughts and desires, compassion and love, willingness to explore more subtle dimensions of reality and to share with others skill, knowledge and wisdom.

Consciousness resonance does not eliminate fuzziness. Fuzziness is an eternal companion to any process of knowing. At the same time, when conscious resonance helps us go beyond the limitations of the fuzziness or succeed in making some problem dissolves, it opens space for new problems to emerge bringing with them new types of fuzziness to puzzle our thoughts and feelings. At any level of consciousness there are phenomena and processes challenging our reasoning, our beliefs and hopes, views and attitudes, aspirations and dreams.

Consciousness resonance fires the bootstrapping algorithm of fuzziness present at a certain level of the individual consciousness.

Consciousness resonance is a creative instant of illumination, a flash of intuition, a sudden understanding of the *truth* of the phenomenon (process, event, actor) on which one's thinking and feeling has been focused. The initial impression is that fuzziness has disappeared entirely, that one has succeeded in experiencing the truth of the studied phenomenon as it is, without using any mental or emotional filter, any borrowed-from-outside knowledge. Yes, the fuzziness has withdrawn itself - has 'pulled itself with its own bootstraps' - from a certain level of one's thinking, experiencing, understanding and knowing - from a certain level of development of one's consciousness; but it has not disappeared forever. It is 'ready to explode' and spread again at the new level of understanding and hence at the new level of development of one's consciousness.

1.10.4 The 'Bootstrapping' Theorem

Consciousness resonance provides both the necessary and sufficient conditions for the fulfilment of the bootstrapping algorithm.

The proof of this proposition lies in the holistic character of the human consciousness - it determines our humanness, it is both the cause and the effect of our human nature; without consciousness we are just animals. The three constituents of the human individuality - body, mind and soul - are three pillars, three powers supporting the individual consciousness and its ability to evolve and grow without limits. The body epitomises the human power to act in the physical world, the mind expresses the power of our thoughts and feelings (and this includes the power of the human heart as a source of our deepest emotions, longings and love), and the soul connects us with the infinity of the human spirit. The resonance between the three human powers represents an apotheosis - the highest peak in the realisation of the creative potential of the human consciousness (at the level of development reached by the individual). Moreover, the resonance triggers a further growth of the individual consciousness, as there are no powers in human nature to support this growth beside those of the human body, of the human mind and heart, soul and spirit.

The growth of our consciousness is at the same time a growth of our capacity to understand (experience, learn and know), and therefore it inevitably affects the fuzziness inherent in this capacity. When our capacity to understand increases, the fuzziness becomes 'thinner' and 'weaker' and folds its dynamics; when there are obstacles on the way of our understanding, the fuzziness becomes 'denser', 'stronger' and expands its dynamics. When the fuzziness 'jumps' from one level of understanding to another, it 'pulls itself by its own bootstraps'. By bringing the mind power to the top of its realisation, consciousness resonance makes it possible for the individual to 'jump' from one level of understanding to another; in this sense, consciousness resonance 'provides' fuzziness with the *necessary* conditions to initiate its bootstrapping. And vice versa, when the fuzziness completes the 'bootstrapping algorithm', it disappears from one level of understanding and appears at another, possible deeper, level (the presence of fuzziness can never be eliminated in an absolute way). The disappearance and re-appearance of fuzziness represent jumps in the human understanding and hence demonstrates spontaneous occurrences of consciousness resonance. In these ways consciousness resonance provides also sufficient conditions for the fulfilment of the bootstrapping algorithm of the fuzziness in the individual understanding.

Consciousness resonance is like a spontaneous coherence occurring with a swarm of 'agents' (insects, birds, ants, neurones, thoughts, feelings, and even people when acting under critical conditions). All the apparent fuzziness of the swarming behaviour suddenly disappears and the 'agents' act in sync and harmony, as if they are *at-one* - one multi-agent entity, one multi-facetted unity, one inseparable whole.

1.10.5 Resonance at Social Level

The term *resonance* has a clear meaning in physics - it is a process of initiating a vibratory response in a receiver that is attuned to an emitter. The emitter is considered as a source of vibrations - they can be periodic, aperiodic or chaotic. In the process of resonance these vibrations 'fire' sympathetic vibrations in the receiver, the magnitude of which is often greater than the magnitude of the vibrations generated by the emitter.

We know about the existence of mechanical, acoustical, electromagnetic, quantum and superstring resonance. The Adaptive Resonance Theory developed by Grossberg (Grossberg, 1988) which is widely applied for modelling human cognitive processes by artificial neural networks, uses resonance between two major neuron fields to explain how these networks learn to recognize, classify and predict patterns and events of the environment both in supervised and unsupervised (without teacher) modes of learning.

Human life crucially depends on the process of resonance. Lehar (Lehar, 1999) argues that the muscle of the heart demonstrates a kind of chaotic resonance, for "the individual cells of the cardiac muscle are each independent oscillators that pulse at their own rhythm when separated from the rest of the tissue in vitro. However when connected to other cells they synchronise with each other to define a single coupled oscillator". In the quoted paper Lehar sees the whole brain as a kind of resonator "whose natural frequency of oscillation as a whole is observed in the global oscillations detected in the electro-encephalogram. This fundamental oscillation sweeping across the whole brain establishes a reference frame or coordinate system in the form of a spatial standing wave, and the higher harmonics on this standing wave represent the spatial percepts of objects perceived in the world, with the phase of those harmonics relative to the fundamental oscillation determining the location of the percept in the perceived world".

Resonance is widely used in descriptions of dynamic interactions at a personal or interpersonal level that are unusually effective, spontaneous, and complete.

"We say that we 'resonate' with an idea or another person when we share an unusually rich set of perceptions that implies to us that we are 'on the same wavelength' - another form of the metaphoric link to physical models. It is a common experience that often is striking in the strength and complexity of shared understanding, and it is associated with successful interactions in pairs and groups of people, and with universally recognized shared experiences. Productivity and creativity are evidently enhanced, and cooperative responses to emergencies and catastrophes seem to be facilitated" (Nelson, 1999).

The world wide web and Internet communication, by making possible instantaneous sharing of people's thoughts, skills and feelings serve as powerful catalyzers for the resonance to occur at the level of society.

The idea of resonance occurring at the level of society closely relates to the concept of *noosphere* - a term coined in 1944 by the Russian academic Vladimir Vernadsky (1863-1945) to describe a new emerging "sphere of intelligence, wherein humanity could employ its evolutionary gifts as a creative collaborative agent of evolution - and where the widening conflict between technosphere and biosphere could be transformed into synergy" (Allen and Nelson, 1986).

The same term noosphere was used in the book "The Future of Man" of Pierre Teilhard de Chardin (published after his death in 1955). He suggested that the Earth in its evolutionary unfolding is growing a new organ of consciousness, called the noosphere. The noosphere is analogous on a planetary level to the evolution of the cerebral cortex in humans. The noosphere is a "planetary thinking network" -- an interlinked system of consciousness and information, a global net of self-awareness, instantaneous feedback, and planetary communication.

In order to plug in the noosphere, the individual needs to discover the 'password'; the role of the password is the level of development of one's own consciousness. Without this password, the noosphere is only a gigantic pile of facts and ideas, pictures and graphics, hypotheses and theories, descriptions and explanations of findings in different branches of human knowing. It is the individual who needs to transform the pile into an integrated whole - a whole which has meaning for this individual, if and only if s/he has succeeded in discovering the 'password'.

How do we discover the password? By concentrating one's physical, mental and spiritual efforts and igniting consciousness resonance. So, it is consciousness resonance - the bootstrapping of the fuzziness of one's individual understanding - that can give the individual access and key for understanding the exciting secrets of the noosphere.

2 Bridging the Study of Complexity with Social Fuzziology

Social fuzziology is inseparable from the new awareness of nonlinear dynamics in every area of human and social life, of living and non-living processes in the universe. The fuzziness of human thoughts, which is at the centre of fuzziology, is humanity's most precious resource in understanding the new world of dynamics waiting to be explored and incorporated into science. It is also the key to the creativity that humanity will need in ever greater abundance to achieve profound and respectful resolutions of the otherwise intractable problems that are building up, and to discover new unsuspected opportunities in surprising and paradoxical places.

At the same time fuzziology does not directly explore complexity as a "non-linear behaviour of systems at the edge of chaos"; the study of this kind of behaviour is the focus of complexity science. So fuzziology is not merely a part of the research in complexity, and nor is complexity reducible to fuzziology. Zadeh's principle of incompatibility makes the connection explicitly: as the complexity of a system increases, human ability to make precise and significant statements about its behaviour diminishes until a threshold is reached beyond which precision and significance (or relevance) become mutually exclusive characteristics.

This principle locates fuzziness in human capacity to make or find meaning in statements (or in the thinking that leads up to and flows from statements), not in the objects themselves. This situates the centre of fuzziology outside the complex systems and the complex world these statements are about. Fuzziology is not a substantive science making truth claims about the world. It is a kind of meta-science concerned with the human capacity to make sense of the world. Yet the fuzziness of human experience and thinking is inseparable from the nature of the world.

The study of fuzziness is of interest not only because of its role, as was thought in traditional science, in obscuring the clarity of crisp truths, but also because of its crucial importance in enabling ways of experiencing - thinking, feeling and acting - that are more powerful in understanding and dealing creatively with the most intractably complex aspects of reality. In social fuzziology there is no dualistic separation of a thinking mind from an unrelated world. The nature of the world considered as an inseparable unity-in-dynamics of the *outward* world - the world around us - and the *inward* world - the world of each individual soul - is the engine of social fuzziology, in general and in every instance. At the same time, it is the engine of our creativity.

2.1 Complementarity and Mutual Enrichment

Zadeh's principle of incompatibility uses the notion of system - a natural or artificial concatenation of elements, which function as a whole. There are of course many phenomena which are not ordered by any 'systematicity' that has yet been discovered, which therefore exist outside the scope of this principle, the world of chaotic phenomena that for many centuries has remained outside science itself. Fuzziology does not propose order in this surrounding chaos, but on the contrary, it recognises the presence of chaos in what had been previously supposed to be the world of order and system. A kind of fuzzy threshold is assumed to exist somewhere at the boundaries of the domain of systems thinking, beyond which its rules and assumptions would cease to apply.

The threshold mentioned in Zadeh's incompatibility principle can be connected with the idea of the 'edge of chaos' of Chris Langton (Langton, 1987). For Langton and before him Prigogine (Prigogine and Stengers, 1984) the zone between order and chaos was highly productive. For Prigogine it was where new more complex forms of order were born, order out of chaos. Zadeh's major contribution to this thinking was to realise that as this boundary or threshold is approached, the old rules of thinking become inapplicable and break down.

Chris Langton described the experience he had which led to his concept 'edge of chaos'. He was scuba diving off the coast of Puerto Rico, and at one point his instructor took him to the edge of the continental shelf, at about 2000 feet:

"It made you realise that all the diving you had been doing, which had certainly seemed adventurous and daring, was really just playing around on the beach. The continental shelves are like puddles compared to 'the Ocean'. Well, life emerged in the oceans, so there you are at the edge, alive and appreciating that enormous fluid nursery. And that's why 'the edge of chaos' carries for me a very similar feeling: because I believe that life also originated at the edge of chaos. So here we are at the edge, alive and appreciating the fact that physics itself should yield up such a nursery" (Waldrop, 1992).

We make three observations arising from this statement.

(1) The concept 'edge of chaos' appeared fuzzy to our mind, and needed to be so. It is 'poetic' rather than 'scientific', though it was created as a result of computer experimentation with cellular automata. It connects with sensuous physical experience at the same time as it makes links between physics and biology.

(2) The concept is therefore simultaneously a major concept in complexity and also a significant example of the creative and productive use of fuzzy thinking to put into the laboratory of social fuzziology.

(3) Thus, fuzziology and the theory of complexity and chaos are not adjacent but related fields of thought, and are more like complementary traditions that flow into the same broad stream of thought. Complexity and chaos theory has so far been exciting and productive as an invisible and unacknowledged branch of fuzziology, renovating the resources of scientific thinking, rather than directly contributing to the stock of scientific knowledge.

Social fuzziology needs the ideas of people who work in the field of complexity and chaos because these ideas are such wonderful examples of holistic thinking at work in the sciences. Complexity and chaos need fuzziology as a framework for them to recognise what it is and what it needs to do so that it can do it more and better, and be understood and valued for what it is.

The danger complexity and chaos face is that they may be dismissed as only a popular fashion, a trendy set of metaphors easy to replace by another set in the minds of the fickle public. This is the essence of a critique by the science writer John Horgan of what he christens 'chaoplexity':

"So far, chaoplexologists have created some potent metaphors: the butterfly effect, fractals, artificial life, the edge of chaos, self-organizied criticality. But they have not told us anything about the world that is both concrete and truly surprising, either in a negative or in a positive sense. They have slightly extended the borders of knowledge in certain areas, and they have more sharply delineated the boundaries of knowledge elsewhere." (Horgan, 1996)

Horgan is right to note that chaos and complexity theories have not yet produced new and startling knowledge about the world. However, he does not note that this was true of Newton also, whose great achievement was to make new sense of what was known. Science has never been merely a set of facts. Even more it has been a productive way of thinking. Fuzziology is a laboratory in which a new way of thinking about the universe is being explored, freed of some key assumptions about good (scientific) thinking that have been dominant for 400 years. Theorists of chaos and complexity are workers in that laboratory, collaborators in a single enterprise so large it is invisible to someone like Horgan.

2.2 'Three Body Problem' in Fuzziology

2.2.1 KAM Theorem

The theorists of chaos and complexity acknowledge significance of the theorem of Kolmagorov-Arnold-Moser (1978) - so called 'KAM theorem' - that deals with the unresolved 3-body problem of Laplacean-Newtonian celestial mechanics - a problem firstly approached by the French mathematician Henri Poincaré (1890). The problem consists in describing mathematically trajectory of an object, the motion of which is influenced simultaneously by the gravitational forces of three bodies. Facing an insurmountable computational difficulty when trying to solve this problem, Poincaré saw possibility of existence of a non-wandering - dynamically stable nonlinear solution of extreme complexity, and thus did the first step into the new mathematical theory of chaos.

KAM theorem asserts that the trajectories studied in classical mechanics are neither completely regular nor completely irregular, but they depend very sensitively on the chosen initial states: tiny fluctuations can make them chaotic.

What has to do the above result with fuzziology?

The dynamics of nature are universal. Its energies and forces act everywhere - in the macrocosm and microcosm, in the so-called non-animated world and in the world of the living organisms, in the human body as well as in the human mind - only the scales of manifestation of nature's dynamics are different.

2.2.2 Dynamics of Fuzziness

We do not know much about the nature and sources of human dynamics, but we know that it is because of them that we exist as thinking and feeling creatures. Our urge to know is a force that can reach great magnitude. While centred in this force, the fuzziness of knowledge is not a static 'object' either; it moves - expands and shrinks, bifurcates or 'explodes' into many streams. It can become dense and almost impenetrable, but it can become also transparent and transpicuous; it can whirl in a vortex together with our turbulent thoughts and emotions, or become quiet when our mind is still.

In this sense, we can speak of dynamics of fuzziness of human knowing, no matter that we can not see and map their phase portrait on the screen of the computer. Our awareness of fuzziness - our ability to see and feel it expanding and amplifying, when we are in captivity of ignorance, or shrinking and almost dissolving, when we transcend its limitations, is the proof that its dynamics exist. The dynamics of fuzziness includes the energies, forces, sources and attractors that produce and sustain it, condense and dilute it, make it move and transform.

However strong the human urge to know, it is oriented towards three main 'bodies' of attraction:
(1) natural environment
(2) society
(3) ourselves

While experiencing the 'forces of gravitation' of the above three bodies operating in parallel, the dynamics of fuzziness ingrained in each of the three streams of knowing may easily become chaotic: extremely sensitive to perturbations in the pulling forces, unpredictable and hard to be controlled. One cannot say whether fuzziness is going to expand or shrink, strengthen or weaken, evolve or transform.

When moving - stretching, folding, increasing, decreasing, 'bifurcating' and reproducing - in unpredictable ways, fuzziness helps us make meaning of our thoughts and feelings, of the thoughts and feelings which others share with us.

2.3 Chaotic Attractors of Meaning

The attractors emerging out of chaotic dynamics of fuzziness of the thoughts and feelings, which 'swarm' in the mental space of an individual, can be interpreted as emergent meanings.

Mathematically, a chaotic or strange attractor is defined as an attracting set with zero measure (that is, a set capable of being enclosed in intervals with an arbitrarily small total length) in an embedding n-dimensional space (called phase space) and has a fractal structure (that is, a structure, which displays self-similarity on all scales of its manifestation). Every attractor has a basin - a limited area in the phase space, where the pulling force of the attractor acts; if an object enters the basin of a chaotic attractor, it starts to skip randomly - in an apparently chaotic way. No matter how chaotic the trajectory of the objects appears for an observer, it stays attached to the attractor.

The cause for a meaning to emerge can be any sign - object of senses, experiential event, word, text, idea, story, etc. - projected on one's mental space. As far as such a projection is only an energy pattern - a kind of whirlpool in the overall fuzziness of the flow of one's thoughts and emotions - it can neither be seen nor touch but only abstractly expressed in an arbitrary small 'attracting set' of mental perception. In this sense, its 'measure' is zero.

Once a certain sign makes sense to an individual, this individual has a capacity to 'zoom' deeper and deeper into the meaning of the sign. Although each level (scale) of the meaning differs from any other level, there is similarity between the levels, as they all relate to the nature of one and the same sign interpreted by one and the same individual. In this sense, the structure of an emergent meaning can be characterised as fractal.

The 'phase space' where meanings emerge is provided by the human mind; we refer to it as a mental space. Human brain is the material embodiment of the mental space.

The mental space is a space of our thoughts and feelings; the meanings that emerge in the mental space of an individual encapsulate the fuzziness of his or her understanding.

Every meaning has its own 'basin' - a zone of relevance (significance, validity, value) - an area in one's mental space that includes the thoughts and feelings supporting one's fuzzy understanding of this meaning. If a thought (or a feeling) is in the basin of an emergent meaning, it serves to convey this meaning. The 'trajectory' of the thought may appear chaotic to an observer, as the thought can 'live' and 'move' in difficult-to-predict and sensitive-to-perturbation verbal or non-verbal contexts. No matter how chaotic this trajectory appears, if it is located in the basin of an emergent meaning, it conveys this meaning.

For example, if an emergent meaning is egoistic, that is, its basin is imbued with excessively selfish thoughts and exaggerated feelings of self-importance, whatever the contexts in which these thoughts and feelings become expressed, they covey the egoistic meaning.

Now we are ready to justify the application of the notion of a chaotic attractor to the meaning that emerges out of the dynamics of fuzziness in one's mental space: each meaning is a specific kind of an attracting set with 'zero measure' and fractal structure. Therefore it resembles a chaotic attractor. But this is not enough. The proof of existence of an emergent chaotic attractor consists in showing the effects that this attractor is able to produce. Is an emergent meaning able to produce effects?

To make meaning of a sign implies an ability to respond to this meaning, to react to it, to undertake corresponding actions. It is the meaning that informs human actions. If there is no meaning behind one's action, the latter is meaningless; it is just a waste of energy. So the 'mission' of an emergent meaning is to produce effects, that is, to transform the fuzziness, which it encapsulates, into action(s).

2.3.1 Emergence of New Meanings

New meanings correspond to new chaotic attractors brought to life out of fuzziness in the mental space of an individual (a group, an organisation, society). According to complexity theory, emergent phenomena are likely to occur in critical zones of the phase space, that is, in zones where some characteristic parameters of their dynamics reach critical values.

In such zones of criticality, chaotic attractors may undergo four types of crises:

- two (or more) chaotic attractors can simultaneously lose their dynamic stability and merge to form a new attractor - a phenomenon known as *attractor-merging crisis*
- one chaotic attractor can become suddenly destroyed - a phenomenon called *boundary crisis*
- one chaotic attractor can dramatically decrease or increase its size - phenomena called *folding interior crisis* or *expanding interior crisis*, respectively
- one chaotic attractor can split into two or more attractors - a phenomenon called *attractor-splitting crisis*.

Similar crisis phenomena may occur with the chaotic attractors of meaning (CAM). The role of the main characteristic parameter for the dynamics of CAM is played by the degree of fuzziness in understanding of its meaning - understanding expressed by an individual or a group of individuals. Example of activities that strongly affect this degree of fuzziness in one's understanding can be various forms of learning, of strengthening individual or group awareness, honing one's intelligence, training the intuitive components of thinking, etc. These kinds of activities 'push' CAM into critical zones where some of the crises described above occur.

From experience one knows that one and the same sign - text, formula, event of life, hyperlink, piece of music, dance, ritual - can be meaningless for some and full of meaning for others. And even for one and the same person, signs perceived initially as meaningless can be transformed into meaningful ones later, and vice versa. This kind of transformation requires again conditions of criticality to arise in the mental space of people; without people's efforts to learn and grow in knowledge, in awareness, intelligence and wisdom, these conditions can hardly come to existence.

The 'crises' which facilitate the emergence of new meanings or destruction of old ones are:

- attractor-merging crisis
- attractor-splitting crisis
- boundary crisis.

The folded form of an interior crisis is non-productive; this type of crisis results in gradual loss of an already attributed meaning - the attractor shrinks in time and disappears without being able to stimulate emergence of any other CAM.

The attractor-merging and attractor-splitting crises directly lead to emergence of new CAM; the boundary crisis brings energy back to the mental space - the energy liberated when the attractor subject to this type of crisis has been destroyed - and thus increases the capacity of one's mental space to conceive with a new CAM.

There are signs, which can trigger so-called 'chain reactions' bringing to life more than one emergent meanings. Often these signs represent words and expressions able to simultaneously 'fire' several creative crises in one's mental space.

In Zen Buddhism, words and expressions used by advanced Zen-masters can trigger *satori* (a state of spiritual climax, enlightenment) in their disciples. The release of this kind of orgasmic power hidden in certain words and expressions resembles a sudden jump of all CAM existing in one's mental space directly into a 'boiling' zone of criticality. Such a jump can be accompanied with emergence of new CAM associated with a higher level of understanding (expansion of consciousness). The new meanings may appear clear (non-fuzzy) from a standpoint of a disciple who has succeeded to go beyond the fuzziness related to the previous (lower) level of understanding; yet, the emergent meanings are bearers of 'new' fuzziness ready to be explored at the new (higher) level of understanding.

2.3.2 Meanings of Dominance and Suppression

In the light of complexity theory, the process of formation of a dominant meaning can be explained through the expanded form of the interior crisis of CAM: a certain meaning attractor expands up to such a degree that it starts to dominate every other meaning emerging in one's mental space. A 'classical' example is the ego-centred meaning attractor that 'dwells' in the mental space of each of us and often dominates this space with one's high opinion about oneself and inflated feeling of pride and superiority to others.

When the egoism is embraced as a doctrine that dominates one's behaviour, it tends to influence any other emergent meaning. In other words, the ego-centred meaning attractor can expand and become influential up to such degree that any other CAM emerges already in the basin of the swollen ego-centred attractor and therefore conveys the same meaning of swelled-headedness and self-conceit.

Similar to the meanings centred in the human egotism are the meaning attractors implanted in one's mental space through propaganda, social brainwash, suppression, violence and other direct or subtle, rough or gentle, simple or sophisticated methods of mind control. These kinds of attractors also act as activators for the expanding interior crises to burst out in the mental space of many individuals at the same time, and thus to impede the emergence of new creative attractors of meaning.

The implanted attractors are even worse than the egoistic ones. The latter are created by us and express the selfishness that is present in human thinking, although the degree of its manifestation differs from an individual to another. The implanted attractors are created by others and imposed upon us as expressions of stronger ego-centred intentions, ambitions and cravings for power. They pursue a huge selfish effect while spreading their ego-centred intents simultaneously through the brains of many people. Today we are witnessing of such an effect harvested by the largest financial corporations in the world franticly involved in the establishment of global economic order.

2.3.3 'Optimal' Degree of Fuzziness

While precision thrives on fixed and locked meanings, fuzziness implies flexibility in interpretation, openness for evolution and potential for transformation of the existing meanings. The characteristic parameter of CAM - the degree of fuzziness in one's understanding of the emergent meanings - needs to be neither too low (as in the case of a precisely defined or imposed meaning) nor too high (as in the case of a watery, wishy-washy meaning that lacks character and substance).

The 'optimal' degree of fuzziness is one that keeps people's urge to know alive and stimulates their further interest in exploring the roots and sources of the fuzziness in their understanding. The 'best' human fuzziness seeks to comprehend itself in the mental space of those who do not lack the determination to reach beyond its limitations.

The criterion for the 'best' fuzziness of an emergent meaning might well be the size and diversity of the spectrum of all insightful opportunities, ideas or/and actions that individuals are able to see, generate or/and undertake in response to this meaning (Dimitrov and Kunchev, 1977). And vice versa, the chance for failure while trying to grasp the fuzziness of an emergent meaning appears directly proportional to the number and variety of opportunities, ideas or/and actions, missed to be seen, generated or/and undertaken in response to this meaning.

2.3.4 Symmetry Breaking and 'Trunsduction'

Although CAM can potentially energise and inform actions, the transition from one's mental space, where CAM abides, to the physical world, where one acts, is not easy - often it requires an extra effort, be it physical, mental, emotional, spiritual.

For example, theoretically we can understand quite well what kind of life style helps us to have better health, yet we do not act according to this understanding. Many addicts know 'clearly' that the addiction can be fatal for them, that 'tomorrow' they will change the dangerous habit in much healthier one, and yet they continue to be trapped in the old CAM.

In the mental space, CAM can function in harmony with the flow of our thoughts and feelings; we can perceive emergent 'virtual' meanings of great plans, dreams and scenarios and to shape them in a perfect symmetry with our current thinking. But at the moment we start to move towards their realisation, the mental symmetry of CAM suddenly breaks.

Spontaneous symmetry breaking occurs every time when the virtual becomes real. The energy, which propels CAM in the virtual reality of one's thoughts and feelings need to be 'transduced' into energy which propels one' actions.

What are the conditions, which stimulates or impede the transformation of one's fuzzy meanings into concrete actions? We come back to this question in the chapter "Understanding Fuzziness of Ourselves".

2.4 Key Propositions in Social Fuzziology: A Complexity Perspective

Given that social life consists of highly complex dynamic entities which are prone to chaotic development, there are a number of propositions to be drawn from general investigations of complex phenomena that are fundamental in social fuzziology as the field concerned with the fuzziness of thinking, behaving and acting in social life.

2.4.1 Interwovenness and Unpredictability

We cannot predict the long-term behaviour of complex dynamic entities even if we were able to construct their precise mathematical descriptions.

This proposition brings to an end the ambitious dreams of social planners and prophets who have looked to models from science and technology to justify their faith in the omnipotent power of human reason to solve all the complex problems of this world. This naïve arrogance has caused a remarkable amount of damage and distortion to the lives of many people over many years.

Social fuzziology has the task of replacing the fallacious crisp predictions of planners based on input as crisp as they are able to achieve, with a fuzzy wisdom, based on fuzzy insights into the dynamic nature of social complexity and a fuzzy sense of possible goals.

The reason we cannot be precise about the social processes over any length of time is because of the complex dynamic interplay of many 'known' and unknown, internal and external, strong and weak human factors, which constantly influence the unfolding complexity of these processes. The degree of interwovenness of a social process' dynamics - forces, energies, substances and forms - is so high that the whole process becomes extremely sensitive to minute perturbations. Even an infinitely small change in the initial conditions under which the social process unfolds (note that every single moment of this complex and dynamic unfolding brings in its own "initial conditions") may result in drastically different future development. The pioneer of the deterministic chaos - Edward Lorentz called this phenomenon "butterfly effect" (Lorentz, 1993).

The butterfly effect has no place in linear systems: in those systems the changes of output variables are proportional to the changes of input variables. Science prefers to work with linear systems - they are easily predictable. For example, we put a bit more sugar in the tea, and the tea become sweeter; if then we add a bit of hot water, the sweetness decreases in an exact proportion to the added water.

Social life is mostly non-linear. By analogy with our cup of tea, a political party might offer a 'tax-sweetener' to the public just before an election. If this were a linear system, politicians should know precisely how many votes will be won by each percentage of tax hand out. In practice, no politician knows this. Sometimes a small amount works. At other times, a larger amount can cause a voter backlash.

The variables, which we use to describe social complexity, have their critical (or threshold) values. If we 'push' a complex system beyond these values, even slightly, huge changes may occur within the system as a whole. In complex situations non-linearity usually leads to unpredictable behaviour. Weather, 'political' as well as meteorological, the stock-market, human and environmental health, national economies, large business or socio-political formations are all examples of unpredictable non-linear systems which need to be understood using fuzzy categories.

2.4.2 Dealing with Complex Causality

Strict linear causality does not apply to social complexity.

Every social process is embedded in the universe, connected in small or large ways with everything else. When the entire universe contributes to the existence of even a 'smallest' thing, how could we precisely distinguish any specific cause-and-effect relationship in this universe? Where everything relates to everything else in a tangled dynamic web of interdependent relationship, how can we trust any linear analysis of this web?

Having pointed this negative truth out to a reluctant world of people who seek to control events for various reasons, good or bad, social fuzziology is then left with the positive task of stirring-up human creativity, helping individuals move beyond the social stereotypes and discover their own ways for navigating the ocean of complexity, instead of dreaming of an exact, safe and tidy linear haven that does not exist.

This discovery gives us a key for understanding the crucial difference between the "classics" of crisp forms of systems science and positivistic sociology and the "heresy" of social fuzziology, the "fuzzy" new science of complexity and chaos. According to the "classics", an effective description of a system can always be built from the descriptions of its separately analysable parts. What is needed is to identify the forces which help keep the parts of that system in a balanced relationship, and to remove the perturbations pushing the system out of equilibrium. Although this approach never works in reality, practitioners continue to use it in pursuit of illusory states of equilibrium of the systems underlying human and social behaviour. As such states never exist, they waste time and resources pursuing mirages.

According to the "heresy", separately analysed components can never create an adequate description of the whole. Positive and negative feedback loops permanently drive the overall behaviour out of equilibrium, towards the edge of chaos, a critical zone between disorder and order, where emergence of new qualitative states takes place, and transformation of the system as a whole occurs.

When we deal with complex social issues, we cannot rely upon the long-term predictions of the experts. Tiny changes in the conditions under which their predictions have been made alter the prediction significantly. We only waste time and energy when trying to force complex system developments in a pre-planned, non-negotiable direction. It is much wiser to learn how to nudge from within the system, how to manage and guide from inside its chaotic dynamics (Goerner, 1994).

Unpredictability of complex behaviour is not an obstacle on the way to grasp this behaviour. On the contrary, by exploring the unpredictable dynamics of non-linear processes, we can gain insights with enormous explanatory power. For instance, this helps us to understand that there are no negligible actions: even randomly chosen and seemingly insignificant actions can lock-in, accelerate and amplify, beyond our ability to control their future directions; because of this effect, complex systems are permanently driven out of equilibrium. In this situation it is better to have 'open minds and hearts' in order to see and feel the subtleties of the system as a whole.

2.4.3 'Fractality' and Wholeness

Reducing does not simplify: interaction of dynamics is important and interaction means integrity and inseparability.

Theorists of chaos and complexity dispense with the idea, foundational in the crisp logic of mainstream science, that the world and what it contains can be analysed into separated parts. If parts simply do not exist as such distinct from the Whole, and the Whole at any scale level of presentation consists only of 'wholes' representing it at lower scale levels, then the distinction between whole and part in any relatively crisp form has been dissolved. If we can never understand how a specific stream of system dynamics works without looking at it as a manifestation of the activity of the whole system, then we have to regard each such stream as both fully a part and fully a whole: a contradiction according to crisp logics, a strain even for many forms of fuzzy logic, since this is not just a matter of being a part up to a degree, (say 0.8) and a whole to a degree, (say 0.2) but being fully a part and fully a whole.

Fractals, discovered by Mandelbrot (Mandelbrot, 1982) are similar patterns repeating themselves at higher orders of dimension. If you magnify a fine area of fractal structure, you get increased information in proportion to the new scale. Thus, the world not only looks different to the observers at different scales, it also measures differently. In every day language, this powerful mathematical insight means that the deeper one's understanding of a complex picture, the more meaningful nuances one can notice in it.

In biology we can see that every cell of the organism bears the unique genetic code of the whole organism, otherwise it would not be able to interact with other cells. Conversely, it is also understood that it is because of the interactive nature of cells that cells have elaborated this code. The cells make the organism function, and at the same time the organism as a whole supports the functioning of each cell.

There is a property of cells discovered by biology that is both a compelling metaphor for the unbroken wholeness of life and at the same time a challenge to fuzzy thinking, to provide appropriately fuzzy categories for biological and social life. Apoptosis is an essential property of all living entities, in terms of which cells are genetically programmed to kill themselves, if not the constant interaction with their neighbours - interaction that keep them alive. No one cell is capable of living in isolation. No one human being either! This phenomenon seems strange to a crisp logic that sees life as the inherent and defining property of all living things, striven for and sometimes lost, but belonging to the living entity and to nothing else. Yet each cell strives for death, not life, unless it is sustained by other cells which would also strive for death unless sustained by others, in a larger living organism whose life is therefore dependent not on the simple 'life' of each component cell, but on a complex mixture of life and death, life as the temporary absence of death.

This unbroken wholeness in which, and through which, we exist is another manifestation of self-reflexivity, which was brillantly captured by Gödel in the proof of his celebrated theorem: it is possible to make true statements within a particular system that cannot be proved by use of the elements and logic of that system. This is because the system under consideration is organically interconnected with some larger system, which by itself is dissolved into another greater than it, and so on.

2.4.4 Far-from-Equilibrium Criticality

In social life - life under far-from-equilibrium conditions - criticality is a typical state.

Everyone who works in a complex social organization, whether in a management or other capacity, needs to leave aside any dreams for 'blissful' steady states of equilibrium in organisational dynamics and learn how to deal with critical states: states which are subject to bifurcations, sharp changes or transitions in a field, which may be single or may form cascades of bifurcations. Although there are periods of relative stability ('punctuated equilibrium'), there will always be critical states and it is these where the practical ability of the practitioner is tested, and not in the steady state of affairs.

Chaotic dynamics are 'ruled' by strange attractors - emerging phenomena with whimsically strange forms (seen when mapped on the 'phase space' - a mathematical space containing all possible states of dynamical systems). Anything outside of the strange attractor is 'folded' towards it, but anything on it is 'stretched' in an unpredictable way – except that one thing is predictable: it always stays on the attractor.

When dealing with complexity of an individual or an organization or a society as a whole, we need to understand what kind of strange attractors are propelling the complex dynamics of this individual, organisation or society. What fields of social activity attract, inspire and concentrate the energy of people and what are the regions where this energy dissipates? Are there any hidden forces responsible for bringing forth specific organisational dynamics? If it happens that the present strange attractors are detrimental for human survival, how can the emergence of new attractors be catalysed which are in favour of personal (or/and organisational) growth?

Chaos is ubiquitous: a chaotic orbit can come arbitrarily close to any point in the phase space. Since chaos can occur on all size scales of human activity, people could use instabilities in order to manipulate the motion of energy in the society on a very large scale. This becomes possible due to the butterfly effect: little changes bring forth significant results. The butterfly effect gives an incredible power to the hands, brains and hearts of practitioners working in various fields of social activity, and in particular in the field of human and environmental health.

Life escapes both the order and the attractors and constantly moves towards the edge of chaos where avalanches of changes never cease to occur. In any 'frozen' region of order or at any strange attractor the dynamics of life would not be able to sufficiently adapt. It is just on the frontier between order and chaos - at the edge of chaos, - where a delicate dynamic balance emerges, impregnated with seeds of innovative transformations. While we cannot be masters of transformation at the edge of chaos, we are not slaves to it. We co-create it! This is an entirely new challenging way of perceiving the role of those who seek to be practitioners of fuzziology - agents of change in the Global Age of Complexity and Chaos.

"*In a recursive, complexly interwoven world, whatever one does propagates outward, returns, recycles and comes back in a completely unpredictable form. We can never fully know to what results our action leads. We take action, the action can have a very potent shaping effect. Then we relax the drive to control and allow the process to unfold - the process learns, shapes and changes itself through all its inseparable components, not under the direction of one of them only. Together with overall changes in the process, we also change, almost unnoticeably, without any strain...*" (Goerner, 1994)

2.4.5 Self-organization and Vorticity

Complex social dynamics give birth to forces of self-organization, which seem to arise spontaneously from 'disordered' conditions, not as simple products of known physical laws.

How can entirely new structures emerge from the multitude of dynamic interactions within a complex system? The concept of *vorticity* sheds some light on this stunning phenomenon.

Vorticity is a characteristic of whirling energy and vortex is a material embodiment of such kind of energy. Examples of vortices are whirlpools, maelstroms, tornadoes, whirlwinds. The larger the vorticity, the stronger the self-organizing force that a vortex produces - a 'sucking force' along the vertical axis of the vortex. Where this force is coming from? It does nor come from outside the vortex; it comes from within it, as a result of suddenly intensified dynamic interactions of masses (of water or air) with different characteristics (say, density, temperature, velocity). The swirling masses are driving the self-organizing force, and vice versa: the self-organizing force sustains the wholeness of the vortex and centers its vorticity. So the emergent vortical structure and the force, which supports it, are inseparably connected.

Similar emergence occurs in our innermost 'space', where the integrity of one's body, mind and soul expresses its limitless power. Is is the space of our thoughts and feelings, ideas and emotions, longings and aspirations, spiritual beliefs and dreams. The whirling motion of these 'immaterial' emanations of our brains and hearts can generate forces of a magnitude comparable with the magnitude of those forces which tornadoes and whirlpools can produce. Only the 'state space' is different; the forces emergent out of tornadoes and whirlpools are in the space outside us (at the visible macro scales of existence), while the forces emergent out of our brains and hearts are in the space inside us (at the hidden micro scales of existence). Same dynamics of nature, different scales of their manifestation!

Heisenberg once said: "*The same regulating forces that have created nature in all its forms are responsible for the structure of our psyche and also for our capacity to think*" (Heisenberg, 2000). Nature does not select a special kind of dynamics for humans and another kind for the rest of the universe; the dynamics of nature - its energies, forces, substances and forms - are ubiquitous. Of course, the ways these dynamics operate at the scale of the macrocosm differ from the ways they act at the scale of the meso- or microcosm. Quite specific are their manifestations at the physical level of the human organism, at the level of our emotions or at the level of our mental activity.

What stimulates the whirling motion of our thoughts and feelings? What injects vorticity in the space of our emotions and ideas, aspirations and dreams? The answer is surprisingly simple: our urge to know, our thirst to explore the world and ourselves, our motivation to go beyond the limits, which the fuzziness puts on our understanding of reality, simultaneously outer and inner.

The wisdom of Vedas considers the outer reality a projection, an imperfect reflection or a symbol of the inner experience. "*Dive into the centre of your own being, into the depth of your consciousness, in which the whole infinite world is contained*!" say repeatedly the sacred writings.

Layers of fuzziness cover the way that leads to the depth of our consciousness. While exploring the nature, sources and dynamics of fuzziness, fuzziology assists us in our endeavour to decrease the denseness of these layers and learn how to transcend them. The impetus for transcending the layers of fuzziness may well emerge out of the vortices of our inner experience.

In the ancient Sanskrit books of yoga, seven major spinal chakras of human body are described, each chakra representing a dormant vortex of vital life energy. Meditation practices and yoga exercises can make the vortices swirl and thus awake forces of tremendous power able not only to invigorate the whole organism, but also to inspire and stimulate the intuitive component of human thinking.

Vortices of immense energy are hidden in human group dynamics. It is a noble challenge for those who seek to facilitate change in any area of social activity to wake them to life. The ability to inspire and ignite imagination, warm hearts and awake hopes and aspirations, radiate joy and stimulate new thoughts and vision, this is what brings forth self-organizing forces out of whirling dynamics in organizations. If this is done with honesty and humility, good will and sincere desire to help and support, the organizational vortices can contribute in liberating people's potential for creativity and growth.

2.4.6 Harmonious Co-evolution

The fuzziness inherent in human experience and knowledge is the only route to a deeper insight into ways of living - thinking and acting creatively - in harmony with the evolving universe.

When aware of the fuzziness imbedded in our understanding and in the ways we make meaning out of our experience, we are prepared to tolerate the opposites that mind generates and harmoniously balance these to such a degree that they cease to cancel each other out, and instead become complementary.

Up to a large degree, social applications of fuzzy logic are oriented towards a search of resilient frameworks - frameworks that are easy to re-shape and modify, - in which the "either/or" approach to contradictory alternatives or concepts can be replaced by a "both/and" relationship of their parallel acceptance. Fuzzy logic has been applied to facilitate this acceptance by assuming that under condition of fuzziness seemingly contradictory concepts can simultaneously be true and false up to a certain degree, so that they can merge into social satisfactory "both/and" statements. Such application of fuzzy logic is crucial for avoiding the paradox of Arrow related to social choice and group decision making (Dimitrov, 1976; Dimitrov, 1983) and also in negotiation where the participating sides are willing to seek consensus (Dimitrov, 1997).

The paradoxical and chaotic nature of social reality causes a great deal of uncertainty and vagueness in the emergence and creation of human decisions. Under conditions of uncertainty and vagueness, when no ultimate answers or best solutions exist, dialogue and search for understanding and consensus between people (at interpersonal level), as well as between the thoughts, emotions, beliefs and aspirations that constantly emerge inside an individual (at intrapersonal level) are vital for managing social complexity.

Fuzzy logic has done a worthwhile job in society by opening the eyes of leaders, and managers to the futility of black-and-white dreams of universality, 'unique solutions' and 'best answers'. It has done a worthwhile job in society by stimulating tolerance to different and contradictory viewpoints and facilitating a switch from contradictory to complementary value-differences.

Fuzziology goes further than fuzzy logic in the search for complementarity and harmony in people's communication and relationships. *By exploring the sources, nature and dynamics of fuzziness imbedded in any set of contradictory concepts, fuzziology looks for new, previously unknown possibilities and decisions.*

As we already mentioned above, the maxim of fuzziology is: "Do not reject, but do not accept either. Go beyond". It is the fuzziness in our knowledge and experience - its sources, nature and dynamics - that constantly challenges us to 'go beyond' what we consider as known.

The 'known' is inevitably partial and fragmentary; it separates us, it makes us compete and fight. The unknown is a bridge to the all-embracing existential wholeness; it unites us, it makes the fuzziness inherent in our endeavours, aspirations, beliefs and dreams resonate with the mysterious fuzziness of the self-propelling drive of our evolving universe. Moreover, it stimulates our urge to explore the harmony of nature and understand how to live according to its eternal laws. Without such an understanding, our every day social and individual behaviour appears animalistic, senseless and destructive.

2.5 Fuzzy Logic and Fuzziology

In this section we explore a paradox about the role of fuzzy logic and its successes in the context of fuzziology. On the one hand fuzzy logic has been a spectacular success story, especially in fields of engineering and applied science, which has put the concept of fuzziness on the map. Its success in these areas of application has contributed to its prestige, leading to a flow of funding to support conferences, research and publications. Yet these successes at the same time have limited the development of the study of fuzziness by itself, as successes lead only to more successes of the same kind, not to new and deeper understandings of the mysteries of fuzziness itself. Unless something happens to reverse this process, fuzzy logic in turn will suffer, shrinking from a paradigm that could change the world of thought into a box of tricks for engineers, alongside other tricks. Over time those tricks will inevitably seem obsolete, replaced by new, more fashionable terms and tricks.

These two tendencies coexist in the present, and they need from social fuzziology a capacity to entertain both parts of the contradiction. On the one hand we need to celebrate fuzzy logic for the work it has done over the past four decades, in an area of intellectual life that was previously thought to be most hostile to its core principles: engineering and applied sciences based on crisp logic and linear models. There must be something fundamentally wrong with the assumptions of traditional science and technology if fuzzy logic works so well. Fuzzy logic has already proved its strong applicative potential.

On the other hand, fuzziness as an essential characteristic of human inquiry did not come from science and engineering. On the contrary it has developed over many thousands of years of human experience and thinking about social, material and spiritual existence. It is not likely that fuzziology has taken all it can and all it needs from this rich tradition in the one scoop, made by Zadeh thirty years ago. Fuzzy logic ought to be an inspiration to fuzziology, if it is taken as a crucial first step into deepening the understanding of fuzziness in all its forms. In its dogmatic or unthinking forms it will be a block to further development. As such, it must be critiqued and resisted.

2.5.1 Fuzziness and 'Softness'

The development of fuzzy systems took place alongside a set of similar tendencies in adjacent fields of inquiry, organised around the concept of 'softness'. Zadeh himself has seen the affinity between fuzzy logic and 'soft' computing, noting that 'the role model for computing with words is the human mind' (Zadeh, 1999). At much the same time when fuzzy logic was introduced, 'soft systems' methods were developed by Checkland and others (Checkland, 1981) out of systems theory because they saw the methods of the previous 'hard' systems approach as unable to deal with complex recursive processes that are common in human systems.

The developments of fuzzy logic and soft systems methods are built around a common discovery that understanding of complex processes and systems can be deeper and more effective by incorporating an organic fuzziness or softness into the basic tools and models of thought. At the same time they have been taken up by powerful and wealthy interests concerned with fuzziness and softness only as means to their own greater profit and control, not because of the insights they hold into the nature and well-being of humanity and the universe. In such a context, the underlying logics of softness and fuzziness are compromised and constrained. It is a primary task of social fuzziology to understand this tendency and provide conceptual means to remedy it.

In the table below, we set out the following differences between FL and possible forms of fuzziology.

Fuzzy Logic	Fuzziology
Studies classes in which there are no sharp boundaries from membership to non-membership and their applications (mainly in the engineering practice).	Studies fuzziness inherent in human experience and knowing, its sources, nature and dynamics, and the ways people deal with it.
Explores a specific type of imprecision associated with the definition of a fuzzy set.	Explores fuzziness as a universal characteristic of human experience and knowing.
Converts human perception into precise symbols for further computer-based processing.	Deals with human perceptions as they are - impregnated with an irreducible fuzziness.
Serves the purposes of control - control over technological processes and machines, engineering systems and robots, natural environment, people, economy and society as a whole.	Serves to sharpen human awareness about the fuzziness of what is considered known, and by exploring its sources, nature and dynamics helps to transcend the limitations it puts on human understanding.
"Accept or reject up to a degree among a given set of possibilities!"	"Don't reject, but don't accept either; go beyond the known!"
Uses fuzzy sets, probabilistic reasoning and soft computing.	Uses insights from the ancient wisdom and nonlinear science.

2.5.2 Role of Fuzziology in Engineering Practice

Engineering for many centuries has been seen as the domain of crisp thinking, in which specific projects are framed with the utmost possible precision according to explicit criteria and exact measurements. Engineering has also always been a practical activity which has required engineers to take account of many other 'fuzzy' factors, if they were to be successful over-all in projects embedded in complex social and material circumstances. So education for engineers consisted of a crisp formal curriculum, and a broad unacknowledged implicit informal curriculum learnt in experience-rooted ways. One of the achievements of fuzzy logic has been to incorporate some of that fuzziness of experiential knowing into the formal and conscious thinking of the profession. Yet the scope of this fuzziness is still constrained and relatively unconscious. There is a need for fuzziology to provide some guidance to help engineers become more conscious of the fuzziness inherent in many of their key professional decisions.

Here are three practical ways for acknowledging this kind of fuzziness:

(1) By using a rule similar to the simple advice of the ancient wisdom for expanding our consciousness: expand the set of variables used in the engineering project, to include as many as possible aspects of the human factors related to this project, and not only technical and economical considerations about the practical efficiency of the project's realization.

Today's advanced techniques of soft computing modelling makes it possible to enrich the engineering models with fuzzy variables more adequately expressing their human (social, ethical, psychological, moral, etc.) dimensions. By experimenting with different ways of expressing fuzziness of what the engineers know (and do not know) about these dimensions, they can test the behaviour of the created models in virtual environments much broader than only technological or economic.

(2) Another way of 'humanizing' an engineering project is by focusing the designers' attention on what remains unknown about the nature of the project and its use in practice. For example, when mutating a gene that is considered responsible for some human disease, one needs to investigate also the relationships of this gene with other ones and how strong these relationships are? Maybe by slightly mutating one gene only, we can crucially influence the behaviour of other closely related genes, and this could have consequences much more severe than the disease that one wants to cure? Similar questions can be asked when introducing a new 'efficient' medical drug. What is our knowledge about the side effects of this drug? How fuzzy it is?

(3) The third way could be by taking an opposite position in relation to the efficiency of the project under development. Instead of collecting only evidences in regard to the worth of this project and how much money it will bring when realized into practice, one may try to collect evidence regarding the potential negative features of the project - particularly, if the latter relates to human and ecological issues of crucial importance.

For example, instead of just repeating how useful the new anti-rocket 'defense' system will be for sustaining the world wide military power o the USA, one can explore social consequences of the realization of this project in a broader human context expanding beyond ambitions for power and dominance. A similar approach can be used when revealing the fuzziness of our knowledge in relation to all kind of global economic projects 'generously' supported by the International Monetary Fund, World Bank, Federal Reserve Bank and other financial corporations which, beside their profit-centred interests, may well have little else in mind.

2.6 Fuzziology and Creativity

Creativity is an essential quality in science as in all other fields of human endeavour. All the great scientists have been highly creative, though the roots and sources of that creativity have not always been well understood. Fuzziology is deeply interested in the conditions of creativity and the kinds and roles of fuzziness for the release of its magic.

What is creativity? When exploring the secrets of creativity in art, Rudolf Steiner wrote: "*We do not create by merely imitating a model; we create by immersing ourselves in the force used by nature to form and create the human being. We shape things as nature does*" (Steiner, 1998).

Translated to terms of fuzziology, this statement identifies the act of creativity with a dynamic and necessarily fuzzy act of fusing inner and outer, subject and object, observer and observed, in an experience of the universal self-organizing force of nature and the way this force works through us and 'shapes' our bodies, our minds, our souls. Such creativity does not attempt to deny or mute the profoundly chaotic and complex nature of human and universal dynamics, but instead learns its profound truths by accepting its affinities.

Another way of putting this is to suppose that creativity is based on a recognition that each human being is a microcosm - a 'fractal', in the terms of chaos theory, which mirrors the microcosm - the universe. Of course, we use the idea of fractality in a different sense than it is used when looking at the marvellous pictures of Mandelbrot's fractals. The shape of the 'human fractal' quite possibly differs from the shape of the 'cosmic fractal'. The similarity is in the ways in which the dynamics that are responsible for the creation, sustenance and destruction of everything that exists, express at the scale of the humans and at the scale of the universe.

2.6.1 Einstein's View on Creativity

One of the most creative scientists of the modern era is undoubtedly Albert Einstein, whose modes of thought have been scrutinised by many in order to understand better what is involved in a high degree of scientific creativity. His own comments suggest the importance of the role of fuzziness in his thought processes:

1. Role of imagination in creating multiple realities in virtual space.

"When I examine myself and my methods of thought I come to the conclusion that the gift of fantasy has meant more to me than my talent for absorbing positive knowledge" (Gardner, 1993).

2. Direct use of fuzzy thoughts, rather than crisp words or concepts.

"The words of the language, as they are written or spoken, do not seem to play any role in my mechanism of thought. The psychic entities which seem to serve as elements in thought are certain signs and more or less clear images which can be "voluntarily" reproduced and combined... The elements are, in my case, of visual and some of muscular type" (ibid., p.105).

3. Multi-modality in modes of representation.

"When Einstein had thought through a problem, he always found it necessary to formulate this subject in as many different ways as possible and to present it so that it would be comprehensible to people accustomed in different modes of thought and with different educational backgrounds" (ibid., p. 105).

Note that this characteristic, which may seem like 'defuzzification', is in practice the opposite, since Einstein's aim is not to reduce his insights to one simpler form, but to represent it in a variety of 'different modes of thought', each of them irreducibly fuzzy in ways that were 'natural' to people inured to those modes of thought, but together adding up to a multi-facetted irreducible fuzziness at the intersection of the different versions.

2.6.2 Inspiration

'Inspiration' is one of the qualities most closely associated with creativity. We can understand 'inspiration' in fuzzy terms taken from theories of chaos and complexity, representing it as the injection of enormous energy into human experiential space – enough energy to facilitate the emergence of new 'strange attractors' in the 'phase space' of the human dynamics in the form of new dynamic clusters of thoughts, feelings and activities, or to evoke changes from one existing attractor to another (Dimitrov and Ebsary, 2000). In the same manner as the resonance between the mind, heart and spirit spontaneously occurs, so does the rapture of inspiration. 'Trying to be inspired' or 'to impose inspiration' is like 'trying to be spontaneous' - it does not work! On the contrary, it creates obstacles for the flash of inspiration to be ignited.

There are many powerful catalysts of inspiration - external (like beautiful scenery, personality, picture, music, poetry) or internal (related to one's individual achievements, self-realization, will power, experience of love, faith, hope). Different catalysts can have different inspiring effects on different individuals. Rarely, a piece of borrowed knowledge can trigger an inspirational sparkle; in most of the situations it remains locked in the field of the mind's activity, as it is used mainly for pursuing goals which are pre-set by one's mind. And everyone knows well that it is impossible to set goals to the heart and the spirit - they hardly obey any instruction coming from the mind.

Only if the knowledge which we 'borrow' from others succeeds in moving our soul, only if it dissolves its boundaries and integrates not only with want we know but also with what we experience, with what we feel, with what we live with and dream for, it could ignite a sparkle of inspiration.

Any genuine spiritual endeavour needs the rupture of inspiration, otherwise it loses sincerity and wilts quickly. Inspiration is needed not only to energize the inner human drive for self-realization and enlightenment, but also to serve as a motivating factor for many people to oppose the injustice and oppression in society.

Inspiration is not a 'logocentric' phenomenon, that is, it is not based on a logically consistent system of thought that claims legitimacy by reference to external and universally truthful propositions. Being the greatest stimulator of human creativity, inspiration requires intermittence (discontinuity) of causality - a quality inherent on the chaotic dynamics; any logical chain of cause-effect melts under lucidity of inspiration. In any rupture of inspiration, both the power of the spirit and the warmth of the heart have a greater weight than the logic of the brain. A posteriori analyses of how inspiration works can possibly reveal some relations of topological similarity of the experiential trajectories of those who have been exposed to the flash of inspiration, rather than similarity of physical causes. Therefore, geometrical mechanisms appear to be suitable for 'mapping' events of inspiration into chaotic experiential dynamics (McNeil and Dimitrov, 1998).

Inspiration evokes forces of creativity; their creative power can grow tremendously, if the synergy between the heart, mind and spirit vibrates with the rhythm of nature. This rare phenomenon can be called a 'double harmonious resonance' (Dimitrrov and Næss, 2001): the heart, mind and spirit resonate not only with one another, but also with the life-supporting rhythm of nature. The level of fuzziness required to understand this goes beyond the fuzziness normally inherent in words, as Einstein noted. For this reason it has become a commonplace to say it is impossible to describe in words the experience which such a creative impetus can bring forth. Maybe the greatest artists, poets, scientists and philosophers of humanity were overwhelmed by such kind of deeply spiritual resonance when creating their masterpieces? Maybe the masterpieces live in eternity because they reflect the eternal dimension of life - its never-ending rhythm, and therefore the flow of time cannot diminish their splendour, their beauty and inspirational power they evoke and will continue to evoke.

2.6.3 The Zen of Creativity

The Zen of creativity pulls fuzziology towards the boundaries of its Logos - the reason that in ancient Greek philosophy was the controlling principle in the whole universe. In the books of Zen, any act of creativity is compared to a leap into a bottomless abyss, with a letting go of all familiar ideas and prejudices. The precipice is the unfathomable depth of our own consciousness, which yawns beyond the narrow circle of our egocentric world of illusions.

The Zen of creativity requires from fuzziology to become conscious of the mysterious working of the creative power of nature trough each of us, through each single entity in the universe, be it animated or non-animated.

With the Zen of creativity we can reveal the mesmerizing fuzziness of our 'progress' in the fields of science and technology - a progress that we make with our brains neglecting the power of our heart and spirit. We cheat ourselves with plans, aims and strategies. We do not hear the whispering voices of our souls because of the noise of our own words. Our restless life takes away our breath, our insatiable desires make our hearts arrhythmic and cause our blood to race through the veins carrying substances that destroy our health. Thus we do not hear the 'sounds of other spheres', do not see the visions described in the ancient scriptures, do not feel the mystery of life... And the eternal creative stream flows past us into the infinite from whence it came.

The Zen of creativity advises us: "In order to hear, you must be silent. In order to see you must close your eyes to the external forms. In order to feel the cosmic rhythm, you must quiet your breath and master your heart. And in order to be carried by the eternal creative stream, you must give up your selfish desires.... Then you could be creative!"

3 Understanding Fuzziness of Ourselves

Dive into the centre of your own being, unto the depth of your consciousness, in which the whole infinite world is contained!
Vedanta-sutra

3.1 Paradox of Mind

In the previous chapter we discussed the ubiquity of nature's dynamics and underlined that there is no reason for nature to select one 'special' engine to support human dynamics and another to support the dynamics of the rest of the world. Neither is there any reason for nature to separate the forces and energies sustaining our physical body from those responsible for emergence of our emotions, thoughts or flashes of inspiration and intuition. It is one and the same engine, which works at all visible and invisible scales of the universe, although the ways in which this engine works at each scale level differ.

It seems paradoxical that the mind, being the 'nearest' to one's innermost nature, constantly strives to externalise - to explore all kinds of external phenomena while neglecting the secrets of its own dynamics. Are the 'offers and demands' of senses so powerful that the mind can not resist their pulls towards what is outside us, towards various objects of external attraction? Or maybe, being a 'Chief Coordinator' of the senses and therefore involved in processing heaps of information, the mind has simply neither time nor will to concentrate upon its own inward dynamics.

Processes like concentration, contemplation and meditation, used in the practice of self-knowing, are self-referential processes: the consciousness relies upon itself to understand itself and uses its own tools to go deeper in exploring its nature and possibly find that unique source and centre of inner serenity, integrity and enlightenment, about which people never stop talking, writing and preaching since time immemorial. Because of their self-referentiality, the processes mentioned above are difficult to grasp and realise at the level of mind only; at that level, self-referentiality often leads to paradoxes, blind alleys, double binds, conundrums. The transcendence or dissolution of these hard-to-digest phenomena needs a special type of awareness - a dynamic awareness developed at a level of consciousness that is different than (and yet vitally connected with) the level of mind. The roots of that level lay in human capacity to experience the living reality holistically and not only through the filters of mind able to produce only series of distorted partial images. The partial mental images serve to aggravate the fuzziness in our understanding of reality rather than to help us transcend the limits that the fuzziness puts on this understanding.

Large is the number of factors, which detract the human mind from achievement of that profound dynamic awareness required for dealing with the self-referential loops of the human thinking. So large and so strong are these factors in their influential power that the mind alone can hardly resist them. Mind is just a 'fractal' of one's consciousness and the consciousness is a fractal of something much greater, something that is impossible to describe with the tools of thinking like thoughts, words, feelings, emotions, beliefs, dreams. Therefore, the application of these tools and their expressions through ever-generated ideas and realisations, decisions and actions, theories and practices appear fuzzy and obscure.

3.2 Journey towards Understanding Ourselves

The fuzziness of our mind's creations does not stop our journey into ourselves. This journey is not an escape from the world or from the human society. On the contrary, with a deeper understanding of ourselves (and the more we discover in us, the more we discover in others and in the world around!), the motivating factors for our resistance against the injustice and oppression in today's world change. They are no more hatred and vengeance, but love of truth and justice, and a genuine concern about the people in society and the creatures in nature that need help.

The journey into ourselves shows how difficult it is for each of us to change those behavioural patterns that one does not like and sincerely wants to abolish, how much will power and effort, persistence and devotion needs the realisation of even a slightest inner change. Patience, sympathy, tolerance and respect to others are qualities, which the inner journey brings up into our relations in society; these qualities are crucial when confronting our views with other fundamentally opposite viewpoints.

The most significant factor for "going beyond" (as socially fuzziology suggests) the destructive friction of confrontation is one's genuine endeavour to understand the reason for it and give examples, through one's own behaviour and actions, of other ways to relate to one another which are much more efficient than confrontation.

The influential strength of the examples that each of us can share with the others, through one's own way of being - living and behaving in society, depends on the depth of our knowledge about ourselves. One can differentiate between three streams of self-knowing (Dimitrov and Ebsary, 2000):

Knowing About the Ideal (KAI). This type of fuzzy knowing may emerge while contemplating the following maieutic question: What kind of identity (personality, inner self) do I want to nourish - grow, realize, reveal - in myself?

Knowing About the Obstacles (KAO) on the way to the ideal. This type of knowing may reward our fuzzy thoughts in relation to another maieutic question: What are the obstacles, both external and internal, that prevent me from realization of myself in harmony with my ideal?

Knowing About the Energy (KAE) supporting one's journey towards understanding oneself. This type of fuzzy knowing may reveal itself while trying to find answers of the third maieutic question: How can I increase and use, as full as possible, my energy potential, strength, willpower, and determination to deal with - surmount, dissolve, transcend - the obstacles on the way to my ideal?

Like the three *gunas* - a Sanskrit name for the three fundamental qualities of human nature, described in the ancient yoga philosophy of Patanjali (Feuerstein, 1990) - the above three streams of knowing are never in equilibrium - they always move so that in each moment a specific fuzzy stream prevails.

If KAI prevails, we often are in a contemplative or merely dreaming mood - either generating ideas, plans, visions and scenarios about our future realisations, or passively imagining fuzzy pictures of ourselves in some desirable ideal states and conditions.

If KAO prevails, we could feel depressed: we might be aware of how difficult it would be to achieve the ideal seen in our fuzzy dreams, plans and visions, and what a large amount of effort, knowledge and vigilance this requires.

If KAE prevails, we are usually in an active and creative mood - we act in order to go beyond the fuzziness of merely contemplating, dreaming or feeling depressed, plus we are keen to undertake actions in order to realize our ideas, plans and dreams about the ideal.

The above three streams interact with one another through various positive and negative feedback loops.

The negative feedback loops between KAE and KAO seems to act against individual self-realization and growth: the less active we are, the more obstacles are accumulated on the way to our ideal; the more obstacles emerge, the less efficient is our activity and this influences negatively our drive to act. There are always thresholds in the number and magnitude of the accumulated and emergent obstacles after which one's motivation for undertaking actions goes to zero.

Most promising for one's realisation in harmony with the ideal's image is the positive feedback between KAI and KAE. Keeping the qualities of the ideal alive in one's mind, heart and spirit stimulates actions towards development of these qualities; the actions sharpen the initial fuzziness of the image and make it appear more real, more close and achievable. When the ideal is felt and experienced as an actually present reality, it can become an effective and irresistible force that stimulates and inspires one's fulfilment in life.

3.3 Human Experiential Space

It is not only in the 'space' of mind where the self-knowledge emerges; the whole life-experience of an individual, with its self-organising fuzziness, participates in the dynamics of the three streams of self-knowing and their interactions. We can imagine the interactive dynamics of these streams as moving (whirling, pulsating, evolving) in a space that is much larger than the space of mind - a space reflected by the whole spectrum of our experience and the ways we live it. This space includes the mental space; we call it "human experiential space" (Dimitrov and Ebsary, 2000).

3.3.1 Characteristics of Human Experiential Space

Human Experiential Space (HES) can be characterised as:

Multidimensional:
An almost infinite number of external and internal interrelated factors affect - stimulate, sustain, impede, or destroy - the interactive dynamics of the streams of human knowing. Although these dynamics results in events and actions that occur in the three-dimensional physical world, in which we move and organize our living, it has not only physical dimensions but also dimensions characterising the wholeness of one's experience - mental, emotional, spiritual. In this sense, HES is a multidimensional space of dynamics specific for each individual's fuzziness of experiencing life.

Chaotic:
One can never exactly describe or predict how the dynamics in HES work and unfold. Events, factors and behaviours that mind considers as insignificant can suddenly bring forth insights of great importance for the process of understanding of oneself (and the world); seemingly simple and routine modes of behaviour can lead to complicated experiential patterns with substantial effects on the fuzziness inherent in this process.

Free from linearity of time:
An hour in the life of a child is an infinitely longer time measure than in the life of an aged person, because the life-rhythm of a child goes at a much faster pace than that of an adult. The less we move, the more we are aware of time. The more actively we are engaged in exploring our inner dynamics through processes like concentration and meditation, the less we notice the passage of time. So, the time span of an event which we experience is essentially fuzzy - not fixed, not linear, and might not be irreversible (as the arrow of time requires) either. (Price, 1996)

Evolving:
The streams of fuzzy knowledge in HES have evolutionary trajectories that mirror an almost infinite and diverse spectrum of ever-occurring experiential events and activities, while affected by the growth of one's individual consciousness - a growth to which these streams of knowledge directly contribute. Human urge for realisation of the inner creative potential - a potential that is in every one of us - is

a force behind the growth of our consciousness and the evolution of our knowledge.

3.3.2 Comparison between Experiential Fuzziness and Fuzziness of Thinking

The fuzziness of HES has different character than the fuzziness of thinking; the latter reflects our ability to understand reality, the former depends on our ability to experience it directly.

Apparently, the factor that strongly influences fuzziness of understanding is our ability for concentration. The higher our ability for mental concentration, that is, the less disturbed the process of our thinking by emergent desires, established attachments or by the diktat of imposed mental patterns, the higher our ability for understanding reality.

One's concentrated mind is at the top of its sharpness and discriminative power when one contemplates events that have already past or ponders future results and decisions. The concentrated mind works perfectly with precisely formulated problems and like to search such problems in the future. What is difficult to remain in the focus of a concentrated mind is one's present experience; the latter is not function of the process of thinking. Our thoughts slide upon the surface of the present while trying to make analogies and comparisons with what has already passed or plans and scenarios about these things which have not yet come to realisation.

The higher our ability for direct experience of life, that is, an experience which is free of comparisons, analogies, classifications, judgements, scenarios, predictions, goals and other 'mindful' operations, the less fuzzy our perception of this experience.

When directly experiencing reality, the mind seems to be switched off, but not our awareness. On the contrary, when the past and future stop pulling us apart, then we can grasp what occurs in the immediacy of the present and be aware of even the slightest motion in our experience. It is like being in an extremely dangerous situation when there is a direct threat for our life; in such a situation the alertness reaches its highest peak and we are conscious about every single detail of reality. There is no time for mindful analyses, conclusions, hypotheses, theories; the person who experiences, the process of experiencing and the experiential events become an inseparable whole. Such a state of maximum alertness (awareness, vigilance) characterises one's direct experience of reality. It seems that the fuzziness transcends in this experience; reality is seen as it is, without any speculations by the reason.

One is able to experience reality directly not only when being in a life-threatening situation, but also in a state of deep communion with nature or in a company with persons whom one really loves, respects and values or when immersed in a genuine spiritual practice, in mediation, etc.

While experiencing reality directly, time seems to disappear.

When there is no time, there is no fuzziness. The feeling of time always implies incompleteness.

After an event happens, one can never repeat again the same experience of it. So unique is each individual's experience that if the message it carries is not understood, one cannot revive it again; the experience moves into the past while leaving behind the fuzziness it has induced in one's mind.

If an event is expected to happen later or tomorrow, it keeps spreading fuzziness in the thoughts and feelings of those involved in its expectation. The proverb says that 'later and tomorrow never happen', but the fuzziness about 'later and tomorrow' constantly reproduces in the mind of the planners, dreamers and prophets.

In the direct experience of *now*, when one becomes at-one with what one undergoes the fuzziness dissolves.

Thinking needs time and therefore the fuzziness is its companion - it increases or decreases when our knowing expands and shrinks in the rhythm of time.

Every act of direct experience resembles the flash of intuition; the latter needs no time - the fuzziness annuls in its illumination, in order to get re-born again in the consequent thoughts and explanations.

3.3.3 'Strange' Attractions in Human Experience

Often the ideals about ourselves that we portray in the streams of KAI - Knowing About the Ideal - have common characteristics which point towards the fuzzy perception of images of an almost god-like ideal creature that is:

- omnipotent
- blissful
- free
- radiating and attracting love
- immortal and healthy
- omniscient.

The mostly spread spiritual traditions of humanity suggest that the humans *are* already the ideal ('God') but in a potential (virtual, implicit, seed-like) form. Otherwise, we would not be able to construct, share and pursue fulfilment of the ideal's qualities.

The qualities of the ideal act as *dynamic attractors* in HES. No matter how diverse one's everyday activity might appear, it seems to be pulled, consciously or unconsciously, towards:

- accumulating and exercising power in various forms, e.g., by acquiring possessions, money, higher social status, influence, prestige, etc.,
- experience of happiness and different kinds of pleasures,
- freedom for opening new possibilities to express one's skill, creativity, intellectual and artistic abilities, talents, dreams, etc.,
- experience of love,
- longevity and health,
- expanding one's knowledge through learning and education, cultural and spiritual enrichment, contemplation, etc.

Power, Pleasure, Freedom, Love, Longevity and *Knowledge* represent six relatively stable dynamic patters - 'strange attractors' in the chaotic space of HES. It is impossible to know exactly how these six attractors manifest in one's experience, behaviour and actions, or to predict precisely the changes they can produce in one's life; in this sense they are fuzzy and even chaotic - unpredictable and sensitive to perturbations occurring in HES. However fuzzy and unpredictable the manifestations of the attractors are, one can easily recognise (just by watching one's actions and behaviour) in the 'basin' of which attractor(s) this behaviour or activity is placed.

The attractors need energy in order to exercise their pulling drives; the energy is supplied by the physical, mental, emotional and spiritual urges and motivating forces of the individuals involved in the creation and experience of these attractors. When the energy supplied to an attractor stops, it shrinks and dissolves into HES.

The attractors in HES remind Jungian archetypes - dynamic structures of thought shared by all humans and imprinted in the human mind like predispositions and habits. Forces similar to various conscious and unconscious urges, which Jung assumed to sustain the archetypes, are responsible for the chaotic dynamics of the attractors in HES (Van Eenwyk, 1997).

3.3.4 Understanding *The Thinker*

There are substantial differences between the qualities of one's ideal and the ways these qualities actually manifest in one's behaviour. As Krishnamurti (Krishnamurti, 1974) explained, together with the process of living - thinking, feeling, acting, behaving, there is another process going in parallel at the inner 'scale' of each of us. This is the process of formation and hardening of the thinker - the person who creates the thoughts and assigns the filters - values, worldviews, opinions, judgements, attitudes, feelings, obsessions, biases, doctrines, theories, philosophies, etc. - through which to see and analyse the events of life.

When the thinker persistently rationalises one's experience - separates, selects, interprets, compares, weighs, evaluates, reshapes, adjusts, adapts, modifies, etc., the experience loses its freshness, its directness, its immediacy. The fuzziness inherent in the thinking process constantly projects on the experience and obstructs its direct perception.

The primary goal of the thinker is to protect its most important creation - the individual's ego: to support the realisation of the ego's urges for survival, for influence and control, recognition and respect. Therefore the root thought for the thinker is always about "I" and "mine".

The deeper the thinker immerses in egotism, the denser the fuzziness that it injects into one's capacity to perceive, experience and understand.

The proof of the above statement lies in the insatiable nature of the egotism. The thinker constantly tries to satisfy the appetite of the ego and hence looks at reality as a place where its main preoccupation is with how to better serve the ego and grab as much as possible from the world in order to respond to the ego's desires and ambitions. While everything is seen through such a narrow egoistic perspective, reality appears partial and distorted in one's mind, and the chance for seeing the large and rich picture of reality in its vibrant wholeness diminishes abruptly. This sustains a particular kind of fuzziness in one's perceptions - fuzziness whose seal is the egotism of the thinker.

Under the influence of an immersed-in-egotism thinker, the attractors in HES transform into egoistic, that is, into attractors, which drag one's mind and body towards realisation of ego-centred behaviours. In the aftermath of this, irresistible feelings of competitiveness and hatred, pride and arrogance, anger and fear, lust and jealousy, anxiety and sorrow spread their dense fuzziness into the individual experience. They pollute our ability to understand, decide and know with ever so heavy, compact and hard-to-mentally-digest fuzziness. Under the veil of this fuzziness, the relations between us cannot but obey the demands of the ego: 'tit for tat' and 'an eye for an eye'.

Life centred entirely in one's ego (where the ego is nothing but a mental creation destined to perish with the death of the individual's body) does not seem meaningful. It hardly makes sense to choose death as a centre point of one's life.

As we mentioned in the first chapter, many of the great ancient thinkers considered human life meaningful if its impetus was centred in something that is not transitory, something that does not decay and die together with the body and the ego. Otherwise they saw the life as a day-by-day survival with a death sentence proclaimed for each individual (in an absolute manner!) at the moment of his or her birth.

Let us involve the thinker in the Socrates' kind of inquiry - a maieutic inquiry into the essence of the human dynamics.

3.4 Maieutic Inquiry into the Essence of Human Dynamics

3.4.1 Existential Whirlpool: The Approach of Fuzziology

(1) *What could be that which does not perish with one's physical death*?

The answer of this question - a question that, together with the other questions considered further in this chapter, relates to the maieutic inquiry of Socrates (and of social fuzziology as well) - is very simple in the framework of the key propositions of fuzziology: what does not perish with one's death must be connected with the focal point, the centre, of the ever-present dynamics of existence.

In the wholeness of the existential dynamics, where everything constantly moves - emerges, sustains, disappears and re-emerges again, must be a centre - an *essence* - that holds all the whirling dynamics in an unbreakable unity (Bohm, 1980).

Unity is sine qua non for the existential dynamics; they change, adapt, evolve and transform from one material embodiment to another only in integrity - interconnection, interrelationship and interdependence. This is also what one of the three main principles of fuzziology - the *Principle of Connectivity-in-Dynamics* – stands for. The roots of this principle are in the famous *panta rei* – 'everything is in flux' – of Heraclitus. More than 2500 years ago, he said: "Upon those who step into the same river different and ever different waters flow down." And it does not matter how different the waters are and how much they change – the flow of the river, the 'flux' – persists in its unity.

As the existential dynamics have always been and will always be, their uniting centre or essence is timeless - non-temporal, permanent, eternal.

However fuzzy the human knowledge about the nature of the all-pervading existential continuum, today's science assumes that the whirling wholeness of the universe, at its macro level, represents a gigantic galactic spiral or whirlpool (vortex). The centre that holds together all the swirling dynamics of this whirlpool must be on its central axis. (The infinity of the existential continuum consists of countless number of galactic whirlpools; the centre of each whirlpool mirrors and relates to the centre of a larger one, in a similar way as the centre of our solar system mirrors and relates to the centre of our galaxy, and the centre of our galaxy - to the centre of a larger 'mega-galaxy', and so on ad infinitum.)

The study of nonlinear vortical dynamics of the whirlpools and tornadoes have shown that at the central axis of any highly energised whirlpool a kind of 'hollowness' forms and sustains - a cavity, an emptiness - with no material filling in it. Such hollowness is theoretically expected along the central axis of the all-embracing existential whirlpool. As the cental axis is without a material substratum, it is free from the effect of time: the emptiness persists along the axis without birth and death, without growth and decay. While empty from material substrata, the centre is impregnated with an immense creative potential, similarly to the potential of the hollowness located along the axis of a whirling black hole - a hollowness full of enormous sucking power.

Being simultaneously voidness (void of substance) *and* plenitude (full of inner energy) *AND* something else which transcends the duality of being full *or* empty (something which is *neither* voidness *nor* plenitude), the hollowness of the centre of the existential whirlpool can stand for a kind of abstraction that is not transient and ephemeral but eternal and essential in the whole existence. We shall refer to it as the *timeless essence* of the all-embracing existential infinitum. The description of the timeless essence both as a logical conjunction (voidness *and* plenitude) AND a logical negation (*neither* voidness *nor* plenitude) is acceptable in the context of the transcendent logic applied by fuzziology.

All embodied-in-matter motions, dynamics and interactions, all animated and non-animated emanations of existence gravitate to the voidness in order to release their energy when undergoing destruction, and to the plenitude in order to be filled with energy and emerge again as manifestations of different forms of matter. This makes the whirlpool pulsate in a unique existential rhythm - a rhythm reflected in the pulsations and cycles of the naturals forms.

The key proposition of social fuzziology about the wholeness and fractality of social complexity is applicable to complexity of the existential whirlpool. While keeping the integrity of the existential dynamics, the impregnated-with-energy hollowness along the central axis of the universal whirlpool endows these dynamics with ability to self-organize into various levels or scales; one can distinguish between the levels of non-animated and animated dynamics, between the levels of plants', animals' and human dynamics. Each level of dynamics reveals sub-levels of self-organisation that share similar features with one another; an analogy can be found in the fractal structures of Mandelbrot which reveal geometrical similarity between their levels (scales), when zooming into them. Each level of dynamics has its own rhythm - a rhythm that reflects the universal rhythm of the whole existential whirlpool.

One can hardly imagine how incredibly great must be the magnitude of the overall self-organizing drive in the universe that serves to support all the levels of the existential dynamics. This drive must act as an omnipotent self-propelling engine (similar to the 'engine' sustaining a gigantic tornado) that feeds with energy the whole universe with its astonishing variety of ever-moving, evolving and transforming phenomena and processes at the innumerable levels of their manifestation.

One can recognise the work of this engine in the blossom of a flower as well as in the waves of the ocean, in the pulsation of a simple cell as well as in the beats of our hearts, in the rhythm of our breathing as well as in the rhythm of the cycles of the solar activity.

(2) *What holds together the integrity of the life dynamics*?

The answer of this question is straightforward: the same self-organising forces that support the unity of the universal existential dynamics are responsible for the unity of the life dynamics. Only the scale where the life forces act is different; they act at the scale of the living species. By zooming into this scale one can unveil the forces, energies and forms that support the unity of the human dynamics.

3.4.2 Timeless Essence: The Vedic Approach

(3) *Can we connect our own dynamics with the timeless essence of the dynamics of the universe*?

They are already connected; the human dynamics form a specific level in the vortical structure of the universal existential dynamics. The challenge is to be aware of this connection and make it work in the span of one's physical life; the earlier in life this happens, the better.

(4) *How can we make the connection with the timeless essence work toward individual growth in consciousness*?

To give an answer to this question is beyond the capacity of the science of our days. Thousands of years ago the answer was found by the creators of the ancient Vedas ('Veda' is the Sanskrit word for 'knowing'), and the validity of this answer has been tested through the life experience of the Indian yogis.

In the wisdom of the Vedas, the whirling energies and forces of the human body form also a whirlpool - a vortex - with an inner hollowness along its cental axis called 'sushumna' in Sanskrit. The axis coincides with the line of the spine column and supports a number of smaller vortices ('chakras' in Sanskrit) centred on it (Judith, 1987). At the low part of the axis is a vortex out of which the urge for self-reproduction emerges; this is the vortex that sustains one's sexual drive - the drive responsible for the continuity of the human kind. The upper part of the axis supports the vortical dynamics of the human mind, which bring impetus to our capacity to know and intuit. Between the lowest and the highest part of the axis, a number of other vortices are centred; they energise one's physical activity, feelings and emotions, as well as the urge to communicate with the others. Just above the middle point of the axis, at the area of the human heart, is the vortex (called 'anahata chakra' in Sanskrit) that brings vigour to our urge for love – the urge to emit feelings of love towards the other human beings and nature, and to feel it when directed to us.

At the basis of the cental axis is located a dormant vortex of enormous potential energy (called 'kundalini' in Sanskrit); this vortex can be activated only if one is able to consciously direct towards it the flow of life-supporting energy (*prana* in Sanskrit) generously supplied by nature. The energy of nature enters the organism mainly by means of breathing, the rhythm of which is consciously controlled and adjusted, so that to increase its efficiency and suit different purposes in the process of activating the inner dynamics. Simple and healthy food, sunshine, yoga exercises, etc. also serve to bring energy to the basic vortex.

Once awakened, the basic vortex starts to pulsate (vibrate) in a rhythm reflecting the rhythm of nature and to release waves of energy upwards, thus 'firing' one after another all the vortices along the central axis. The basic vortex serves as an intermediate link between the inner dynamics of the activated human vortices and the outer dynamics of nature, through which the universal existential vortex expresses its infinite power. Under constant guidance of the individual consciousness and vital support of natural energy, the excited vortices increase their level of energy and tend to synchronise their rhythms.

The higher the energy levels of the vortices, the greater the magnitude of the creative forces they are able to produce. The produced forces have a triple function in the life of the individual; they stimulate the individual efforts:

- to grow in consciousness;
- to search for harmonious relationship with nature and with the others;
- to stay consciously connected with the pulse of the existential vortex and its timeless essence.

So, the answer of the maieutic question "How to make the connection with the timeless essence work for the individual growth in consciousness?" in the context of the ancient Vedas is simple and laconic: by mastering the vortices - the chakras - of one's inner individual dynamics.

Although the activation of all the vortices is required for the individual to stay connected with the timeless existential essence, the Vedas emphasise on the importance of the heart chakra responsible for one's expression and experience of love, and the mind chakras, responsible for one's understanding and knowing. The creative forces released by these chakras are the most powerful catalysts for realisation of the individual potential.

3.4.3 On Practical Realisation of the Vedic Approach

The Vedas describe highly efficient ancient methods and practices of various types of yoga, designed to help the practitioners stay connected with the 'engine of the universe' by mastering the vortices of their inner dynamics.

Three ancient methods - relaxation, concentration and mediation - play crucial role for the practical realisation of the Vedic approach

In a state of deep relaxation, when not only the physical bodies of the practitioners are free of stress and tension but also their minds and hearts are calm and peaceful, they can experience the energising effect of the 'emptiness' at the centre of their inner dynamics. After such an experience, one feels like being born anew, charged with natural energy, the effect of which is similar to the effect of inspiration. Because of this re-vitalising effect, today's science agrees that the relaxation positively influences the health of the human bodies, brings peace and calmness to the otherwise stressful and turbulent human minds and fills the hearts with warm feelings of kindness and good will.

It is in the state of relaxation, when the thinker stops serving the ego and becomes silent; then the obstacles, which the ego unconsciously or consciously creates on the flow of the natural energy, gradually dissolve. When there are no obstacles, the flow of energy is ready to activate the inner dynamics; what is needed is concentration of the mind attention on the vortices - chakras - that these dynamics form.

Concentration seems to be opposite to the relaxation. The energy, which freely move through every single cell of the human body, is now focused on each chakra, starting with the basic one and moving upwards along the spine column. This concentrated energy activates the chakras and makes them release forces of large magnitude. The climax is reached when the heart and the mind chakras are fully awakened, their creative forces strengthen and amplify the overall effect of the mind concentration of the practitioner.

Mediation helps practitioners 'listen' to the rhythm of the all-pervading existential vortex, adjust the pulsations of the chakras to this rhythm, and make the practitioners experience that blissful oceanic feeling of union ('yoga' in Sanskrit means union) with the timeless essence. Meditation combines the effect of relaxation (peace, calmness and tranquillity) with the effect of concentration (sharpening the flow of energy and its 'firing' effects). In Vedas the stage of mediation is considered deeply spiritual; the life-supporting rhythm of the universe is often characterised as divine.

Here is a typical text from the ancient Yoga Sutras of Patanjali - the first ever written book to explain the principles, methods and practices of yoga:

"I am not this body, not these senses, not this mind and not even this intellect. No, I am none of these things! These things are ever changing and impermanent. These things have a beginning and an end. I am not bound by these things, and I am not deceived by them. I am the essence of existence, which has no birth, no growth, no decay, no disease, no death. Therefore, the sword cannot cut me asunder! The spear cannot pierce me through! The wind cannot dry me! The water cannot wet me. The fire cannot burn me! The sun cannot scorch me! I am unborn, immortal, immovable, unchangeable, all pervading and infinite. I am the spiritual power of the essence, which is ever by itself, one without a second!"

3.4.4 Compatibility between the Vedic Approach and the Approach of Fuzziology

The Vedic approach for exploring human dynamics is consistent with the approach of fuzziology. The practical realisations of the both approaches aims at strengthening (energising, amplifying) our ability to grow in wisdom: to understand and know, to search and practise ways for developing and honing the intuitive component of our thinking. For the approach of fuzziology, the emphasis is on firing the bootstrapping algorithm in human understanding, which makes fuzziness withdraws itself from the levels of an ever-deepening process of understanding and expanding the consciousness. For the Vedic approach, the emphasis is on firing the chakras at each level of the human vortex. The ascending journey of one's consciousness through the activated chakra at each level of the human vortex is, at the same time, a process of deepening of one's understanding, expanding the individual consciousness and therefore, self-withdrawal of fuzziness. In this sense, the practical realisation of the Vedic approach represents an efficient realisation of the bootrsrapping algorithm of fuzziology.

Both approaches are essentially dynamic. The key proposition of social fuzziology with regard to the wholeness of the human dynamics, their ability to self-organise into multi-layered 'fractal' structures, the vorticity of these structures - all these propositions are valid in the Vedic model as well.

The vortical structure of the Vedic model can be characterised as fractal; the whole vortex of the human body is nested in (and mirrors) the gigantic existential vortex of the universe, the body's chakras are self-organised into a holistic multi-level structure. The chakras are inseparably connected with each other, and yet the chakras are nested into one another: every activated chakra can fully affect the functioning of the chakras at the levels below, and have only a partial influence on the upper-level chakras. The highest chakra - the vortex of the mind dynamics, which is responsible for the development of our intuition, direct experience and knowing, can affect all the other chakras. The 'firing' of each chakra cannot occur before the activation of the chakras at the levels below.

The body-mind-soul resonance considered in fuzziology as a stimulator of the bootstrapping pull of the fuzziness demonstrates its practical realisation in the Vedic mediation that synchronises (brings in resonance) the rhythm of the activated chakras. Although in the process of mediation the thoughts and feelings seems to float fuzzy in the mental space of the individual, that is, without any crisp content or direction, their fuzziness is impregnated with enormous creative energy - the same energy that permeates the vortex of the inner individual dynamics. And it is this energy that makes the fuzziness related to one's ego (with all its degrading aspects rooted in ignorance, doubt, hatred, fear, pride, envy, lust, jealousy, etc.) to shrink and withdraw itself. The consciousness of an advanced meditator is described in the Vedic texts as crystal clear - a kind of consciousness that almost perfectly mirrors the all-embracing consciousness of the universe: eternal, omniscient and omnipotent.

3.4.5 The Most Difficult Maieutic Question

By combining the two approaches - the Vedic approach and the approach of fuzziology - we are ready to answer the most difficult and 'intimate' maieutic question:

(5) *When the body dies, and hence the mind and the heart stop functioning, what happens with the individual dynamics?*

This is similar to what happens with the leaves of a tree: the leaves fall and die but the dynamics supporting the life of the tree become stronger, because of the vital help that the leaves have provided for the growth of the tree when they were alive. After the leaves release their energy into the 'hollowness' passing through the centre of the tree level of the plant scale of the vortex of the universal existential dynamics, the 'hollowness' become richer in energy, and increases its creative potential.

The body, the heart and the brain die but the human level of the existential vortex grows in its creative potential, because every human helps, up to the degree of development of his or her consciousness, for the growth of this potential during one's life. And we know that there are not only physical dimensions of the human level of the universal dynamics; there are subtle dimensions beyond the grasp of the thoughts born out of our mortal minds and of the feelings born out of our mortal hearts. The timeless essence that passes through the human scale of the existential vortex becomes richer in its creative potential after the death of the individual; so do the subtle dimensions of that scale. If we are aware about these subtle dimensions when alive and contribute consciously to their development and strengthening (say, by mastering the vortices of our inner dynamics, as the wisdom of the ancient Vedas suggests), physical death would be powerless to destroy our connection with them. So, individual dynamics may continue exercise forever their self-organising power along the subtle dimensions of the human scale in the eternal vortex of the existential dynamics.

According to the Vedic approach, the question is not whether we are finite or infinite, mortal or immortal, but whether we identify ourselves with the infinite and imperishable or with the finite and ephemeral. The ego is finite and ephemeral, the timeless essence is infinite and imperishable.

Does it not seem 'rational' for the thinker to try to identify itself with the infinite and imperishable essence of existence and not with the ego that is destined to die? The thinker encapsulates the power of the individual to inquire - explore, understand and know, so why not to apply this power to inquire into possibilities for going beyond the limits of the physical life before the source of this power dry up?

3.5 Empowering *The Thinker*

If the thinker undertakes the first necessary step: to see the individual dynamics as exponents of something much more important than the ego, something which is beyond decay and death, then there is a chance for its thinking ability to be involved in a 'bootstrapping algorithm' that may result in its transformation from a slave to a master of the ego. Otherwise the thinker will continue to 'know more and more about less and less', that is, to do what the most part of us is doing - wasting our precious time and energy in activities aimed only at ephemeral achievements.

As every bootstrapping algorithm needs to be run from within the process dynamics, a change in the orientation of the thinker's reasoning power is required.

Instead of directing this power outwards in order to show the world how clever the individual is - how rich, great, influential, brave, proud, good, spiritual, wise, polite, intelligent, etc - the thinker needs to orientate it inwards, and thus to support the bootstrapping of its own fuzziness. So, the second step necessary for the growth of the thinker's inner capacity to understand, learn and know consists in interiorisation of the thinker.

Interiorisaton does not mean turning off thinker's attention from the world around, from the vital web of relationships with the natural environment and the other individuals; on the contrary, interiorisaton means honing thinker's awareness of what happens in the outer reality. But now the events occurring in the outer world are projected inwards and each event from the outside reality is seen as revealing something essential about the inner nature of the individual, about the purpose for which this individual has come into a physical life. Each event starts to be a lesson that offers experiential knowledge for the growth of the individual's ability to understand and know.

(6) *Why are outer events - phenomena, processes, encounters with people, various activities in which we are involved - bearers of features of our inner nature, of the purpose of each individual's life*?
This is again a typical maieutic question from the Socrates' treasury of fuzziology.

Once the thinker accepts the connectivity-in-dynamics principle of fuzziology, everything that happens in our outer world relates to our inner one, but whether one can see or not this relation depends on the level of development of one's consciousness. This is also the message contained in the possibility theorem of fuzziology. So, when the bootstrapping of the fuzziness in the thinker's reasoning is activated, that is, when the self-propelled (*autocatalytic*) development of one's thinking capacity advances, one can better see the connection between the outer and the inner, between the universal and the individual. Eventually the thinker understands that the outer and the inner are two scales of manifestation of one and the same existential dynamics. Now a new challenge arises - to make this understanding work in the every day life of the individual.

(7) *What does it mean to make the understanding of the unity between the inner and the outer dynamics work in the every day life of the individual*?

The answer of this question - the seventh maieutic question in this chapter - contains seven points. To make the unity of the inner and outer dynamics work means to succeed in:

- transcending the limitations of the ego;
- centring one's life in the timeless essence of existence;
- understanding the meaning and the purpose of one's life;
- living in a healthy way - in harmony with the laws of nature, with the inner urge for self-realisation, with the meaning and the purpose of one's life;
- growing in wisdom;
- helping others to live healthy and grow in wisdom;
- developing ability to experience, understand and master the spiritual dimensions of the human existence.

The realisation of each of the seven points depends not so much on the unpredictable interplay of the external - outer - factors, but rather on the powers of one's own mind, heart and soul. Initially, the above seven points may appear too fuzzy for the thinker, but one needs to remember the self-organising potential of the fuzziness inherent in the thinker's capacity for understanding. When this capacity grows, it may fire a consciousness resonance that makes fuzziness shrink and withdraw its dynamics.

In order to facilitate the bootstrapping algorithm of the fuzziness, the thinker must watch its dynamics (reflected in the flow of continually emergent thoughts and feelings) quietly, in a relaxed manner, without hurrying to judge or criticise, to praise or blame the individual or other external factors. What is needed is to face honestly the individual's urges and desires, as they appear and dissolve, and not to fight with them, but not to be their slave either - just a serene witness of their interplay. Such an honest witnessing helps the thinker better understand the drives of the ego; each step in this understanding is at the same time a step of moving out of the basin of attraction of the ego. With the decrease of the influence of the ego, the fuzziness of the thinker starts to lose one of the most powerful channels of its support - the channel of the egoistic desires and ambitions of the individual; this results in preparedness for the start of the bootstrapping algorithm.

The third necessary step in the realisation of the thinker's power is to make the thinker act in its space - the mental space of the individual. What the thinker can do is to seed emergence of new 'attractors of meaning' in this space - attractors which strongly pull the individual thoughts towards meanings free from the influence of the ego. What are these meanings? Meanings that evoke one's genuine experience of selflessness, generosity, forgiveness, compassion and readiness to help and care for others.

When these kinds of meanings stop to be only mental constructs but become also practical realisations, then their effects spread in the human experiential space. The six attractors in this space - the attractors of freedom and love, power and knowledge, pleasure and longevity - then move out from the orbit of the thinker's ego towards cultivation of the qualities of the human ideal: freedom from restrictions and obstacles on the way to grow in spirit and radiate unconditional love, limitless potential for realisation of the creative power of the wisdom, and blissful experience of being at-one with the eternal essence of existence.

4 Understanding Fuzziness of Society

According to the proverb, two minds are more powerful than one, so one could expect that the collective mind of society (as reflected, for example, in all the books written up to now plus the world wide web) has better ability to understand and know than an individual mind. Unfortunately, this is not the case.

4.1 Infinite Life of Society versus Finite Lives of Individuals

The collective mind is like an ever-growing encyclopedia of facts, theories, hypotheses, doctrines, pictures, etc.; the ability to see them in their connectedness, to make sense of this connectedness and to use it for one's holistic growth in mind and spirit, depends on the individual alone. It is the individuals, and not the society, who reveal facts, formulate hypotheses, create theories, make pictures, establish dogmas and doctrines. Society is a web of dynamic interactions and interrelationships of people and their natural and artificial worlds. It has emerged out of the joint efforts of the individuals to survive on this planet, and has no separate existence from the dynamics of people's relationships and interactions.

The individual dies, but the human society seems to continue forever, because the dynamics which it expresses go forever; one civilisation falls, another emerges, expands and then shrinks to give place for the next. The cycles of ups and downs in the deathless life of society are manifestations of the self-organising capacity of the social dynamics. Before the life on our planet disappears (due to the physical death of our sun or cosmic cataclysms or because of our ecological ignorance), some group of humans or human-like robots will possibly move to live somewhere else in the cosmos and thus initiate another society.

The infinite duration of society contrasts sharply with the finite duration of each individual life; this essential difference reflects on the processes of evolution of the individual and society.

History keeps alive the memory of wise people from the past, of philosophers and spiritual leaders who, in their lifetime, succeeded to see far beyond the fuzziness in which the greatest part of society abided. It seems that they moved quickly through the stages of the evolution of their minds and souls, realising that there is no time to waste - the individual life is extremely short. But society moves very slowly on the way to wisdom, so slowly that it is difficult even to notice whether it really moves or not. And why to hurry? There is plenty of time for the next generations to 'grow', if they decide to. And where to hurry to, what to attain? Wisdom? Enlightenment? Spiritual growth? What do these words mean for the web of human interactions that we call society? Nothing, absolutely nothing! Some individuals may fill these words with meaning, but for the majority of people they are empty of content.

It looks as though for the majority of us the most meaningful endeavour appears to be satisfaction of our animalistic urges for food, shelter and pleasures. This was expressed a long time ago by the famous Roman formula *Panem et Circenses* (bread and circus games) - the favourite formula for the Roman leaders who wanted to keep the allegiance of the masses; this formula is perfectly applicable for the majority in any 'civilised' society.

Many generations have lived and died after the first tribal societies, but the essential features of the social behaviour persist without change. All the societies have been involved in bloody wars, in spending enormous time and efforts to create more and more sophisticated tools for:

- killing people
- intensive exploitation of natural resources, natural fauna and flora
- wild strife towards power and possession
- merciless punishment of rebels who dare break the restrictions assigned by those in power and who spread 'heresies' about possibilities of social changes
- violent suppression of women, children, poor and disabled people
- savage pursuit of primitive animalistic pleasures.

4.2 Collective Mind: A Pile of Fragmented Knowledge

When one generation hands the baton to the next, it passes its knowledge and skills, religions and arts. It passes also its delusions and fuzziness inherent in its theories, philosophies, doctrines, customs, habits, dreams, expectations. What is hard to pass is how to understand and realise the perennial wisdom, as it carries indispensable experiential dimensions, which cannot be transferred or borrowed from books, prophets and experts; their only verification is through one's individual experience, through living in these dimensions.

The Danish philosopher S. Kirkegaard (1813-1856) once said: "That which is genuinely human, no generation learns from the foregoing" (Westphal, 1996). The genuinely human needs to be rediscovered, experienced and understood by the individuals of each generation; nobody can pour this into one's head. It is easy, on the other hand, to pour in social mediocrity - 'they-say' views and beliefs, prejudices and biases, norms and rules, classification schemes and labels - all that impedes one's own knowing of what is genuinely human.

Similar to what is genuinely human are the words of wisdom; they can evoke fuzzy thoughts and feelings, but the truth of these thoughts and feelings is revealed when this resonates with one's own experience and reaches the heart and soul of the individual. Otherwise the words remain just words, and the more we repeat them, the more senseless they sound. "Truth repeated is no longer truth; it becomes truth again only when it has been realised by the speaker as an immediate experience" (Huxley, 1990).

The bootstrapping algorithm of the fuzziness inherent in people's understanding cannot happen at the level of society. There is simply no place to 'install' and realise this algorithm, as the consciousness resonance that is responsible for its completion is a resonance of one's individual body, mind and soul. Society has neither a suffering body nor a soaring soul, and its collective mind is just a pile of fragmented human knowledge. As we already mentioned in the first chapter, the 'password' for making sense of this pile can be discovered at individual level, when the individual consciousness succeeds to go beyond the fuzziness of one's understanding of reality.

4.3 From Understanding Ourselves to Understanding Society

From the standpoint of social fuzziology, the process of understanding society goes in parallel with the process of understanding ourselves, as all the qualities which we have at individual level, reverberate at the level of society. Any process of understanding unfolds within us; this is expressed through the basic postulate of fuzziology. So, while unveiling the fuzziness from the process of understanding ourselves, we unveil the fuzziness from the process of understanding society. Plus, we are much closer to ourselves - to the thinkers in our brains, feelers in our hearts, and soarers in our souls - than to any other member of society.

Of course, the forms, the ways and the medium of expression of the individual qualities in each human being are different, but their essence is similar. In a way, we are relatives to one another: long time in the past, existential dynamics made it possible for the 'primordial soup' to appear on our planet and give birth to the first proteins, the first cells and simplest organisms. After innumerable loops of the spiral of life - a spiral full of spontaneous emergence, bifurcations and transformations - the humans appeared. In this way we all were connected as potentialities in the common fountain of life on our planet, and we continue to be connected as actual realisations through the common life-supporting conditions on the earth. We constantly contribute to the changes of these conditions and they change us, so that we co-adapt and co-evolve as an inseparable unity.

This stream of thoughts makes us aware that the qualities expressed in the characters and behaviours of the other individuals in society reflect our own qualities. Moreover, the sole ability to see a certain feature displayed in another person is because we already have this feature in ourselves, otherwise we simply would not be able to recognise it, to response to it and to react with criticism or approval, hatred or love, repulsion or attachment.

We easily see the egotism and greed manifested in the behaviour of the others, because we ourselves are full of it. Since the earliest years of our conscious life, we know that to be rich is good and to be poor is bad (no matter that we read the opposite in the Bible, Koran or in the Buddhist texts); so we never cease competing with others for more money, power and pleasures. The competition perpetuates our egoistic drives and strengthens our egos; almost everybody wants to win, to become the first, to be praised, to be at the top.

By amplifying the egoistic traits of human nature, particularly those related to accumulation and realisation of power, society increases fuzziness in our perception of reality.

In societies where the egotism is put on the highest social pedestal, everybody must bear an assigned ego-identity and have a specific ego-status. It is difficult to be authentic and genuinely human when each person is sealed with socially approved labels, as these labels make us perform roles and carry masks. The role performance impedes the individual endeavour to discover and understand his or her authentic self - the inner identity, the centre that is free from the socially endorsed pull of the ego; the further from the centre, the fuzzier one's perception of reality.

The competitive egoistic dynamics of society permanently produce stress in our lives; when there is a stress, the fuzziness increases.

Society is hostile to individuals who seek to trigger radical social changes; its rulers do not like changes, as the changes may affect their established status. In different historical periods, society killed Socrates and the students in the school of Pythagoras, burned Joan of Arc and Giordano Bruno, nailed Jesus, shot Gandhi and Martin Luther King, and millions of other people (with known or unknown names) who honestly wanted and worked in order to show society new ways for its development - ways beyond the manipulations and suppressions of the rulers. Those who were not killed gained enmity from the majority of the people, as the governors knew very well how to excite and stupefy the crowd. "Wherever there is a crowd there is untruth" (Kirkegaard, 1958), and wherever is untruth there is impenetrable fuzziness in human understanding.

We referred to such acts of animosity and cruelty towards the bearers of new understanding of society as *Gorbachev Phenomena* (Dimitrov et al, 1996). Gorbachev, more than any other person in recent history, personifies the effect described. Arguably, he almost single-handedly brought about changes that amplified into historical events of great significance. For his interventions, he firstly earned the Nobel Prize followed by the unforgiving enmity of contemporary Russian society (at the first democratic elections for President of Russia, soon after the collapse of the Soviet Union, Gorbachev attracted only 0.5 % of the primary vote).

4.4 Lack of Wisdom in Society

Is our society wiser than the societies and civilisations that lived before us?

- Is it wise to spend an astronomic amount of money for the production of more and more advanced tools in order to kill each other?

- Is it wise to be constantly involved in wars and to suppress people (and nations) who do not want to obey the orders of those who are thirsty to exercise their destructive military or economic power?

- Is it wise to commit horrible ecological suicide year after year by destroying irreversibly the natural environment and refusing to take even a small step (such as signing a joint International agreement), in order to initiate restrictions on the spread of our fatal socio-ecological madness?

- Is it wise to extinguish the indigenous people (in the same way as the white 'pioneers' did with all the aborigines in Tasmania), or to brutally suppress and force them adopt the 'advanced culture' of the conquerors?

- Is it wise to keep for years those who seek asylum in camps similar to the concentration camps of the Nazis, suppress their protests and impassively watch how they kill themselves, as is currently done in Australia?

- Is it wise to follow politics of discrimination on the basis of race, nationality, culture, gender, political opinions or spiritual beliefs?

- Is it wise to force humanity into the establishment of a global economic order which is threatening to turn the largest part of the world population into economic slaves of a handful monstrously rich corporations, financial institutions, organisations and individuals?

- Is it wise to interfere with the finest and most complex fabric of the human genetic structure aiming at partial effects while putting at risk the integrity and the evolving potential of the whole structure?

This is what our society does. Such society is not wise. It multiplies and amplifies the worst in the human nature - the egoistic urges of the 'rational' thinkers abiding in our minds.

The practical endeavour of social fuzziology is to show the delusive fuzziness that these 'rational' thinkers spread into the life of the individuals and society, and thus to help people understand the nature of this fuzziness.

Further in this chapter we focus on the fuzziness related to the frantic establishment of global economic order in today's society and its impacts on the human education and health.

4.5 Towards Establishment of Global Empire

Today we are witnessing emergence and establishment of an all-encompassing capitalist order - a global empire, defined by Hardt and Negri as "a decentred and deterritorialising apparatus of rule that progressively incorporates the entire global realm (of society) within its open, expanding frontiers" (Hardt and Negri, 2000). A striking symptom of this establishment is the increasing tendency for the national governments to act as an armed militia for the biggest multinational corporations and financial institutions. It is easy to see that the democratic elections in the 'developed' countries of the empire have turned into a 'choice' usually between two major parties with only superficial differences in their platforms; the both derive from the same hegemonic ideology, carefully packaged in media campaigns as presenting opposite points of view.

On the way to the establishment of the empire, the economic gap between those who count in society and those who have no power to change anything has become so large and deep, that the people from the both sides of the gap resemble different species. They have totally different life styles and conditions in which to realise their potential, to care for their health and education, to spend their leisure time.

Extremely precious features of human identity rooted in the treasure of the ethnic culture and indigenous spirituality all around the world disappear brutally smashed by the global consumption-oriented stereotypes of today's intensive and aggressive implantation of anti-human economic, social and cultural models.

4.5.1. Hegemonic Ideology

The hegemonic ideology of the global empire is nothing but a large-scale manifestation of the same egoistic patterns embedded in the human individuality.

In the empire, Dostoevsky's prophecy about the future social order expressed through the prediction of one of the characters of his novel "Demons", Shigyalov, becomes a reality; he sees "the division of mankind in two unequal parts. One tenth is granted freedom of person and unlimited rights over the remaining nine tenths. These must lose their person and turn into something like a herd, and in unlimited obedience, through a series of regenerations, attain to primeval innocence, something like the primeval paradise - though by the way, they will have to work" (Dostoevsky, 2000).

As it is in Dostoevsky's prophecy, so it becomes in our society. The greatest part of it ether lives in poverty or work like crazy to pay back loans and debts taken in order to have places to live and reproduce. The sizes of the loans, debts and the increase of the percentages are designed in such a way that to make impossible their payment back in one's lifetime, no matter how intensively one works.

The lack of work torments a huge number of young people full of unrealised energy and potential to create, makes them psychologically sick, push them towards drugs, crimes and suicide. The 'freely' elected governors hold charismatic speeches against the use of drugs, but no serious actions follow their speeches. It looks like the drugs are perfect tools to stupefy the young people and make them unable understand the roots of the social injustice and its main perpetrator - the richest elite that carries major responsibility for the economic, social and environmental agony of the young generation. Cheap chemical drugs keep the mouths of the young people shut, their physical bodies helpless, their brains suicidally ecstatic, and eventually kill them.

Alcoholics and gamblers are treated similarly in today's world - they do not present any threat for the establishment. On the contrary, they bring a lot of money into the empire through the gigantic octopus of pubs and casinos. There is not much difference between the addicts and the prisoners of the empire - the addicts are prisoners of their psyche destructed possibly due to a constant fight and stress to live normally in a distorted, abnormal society. The greater the number of addicts, the greater the number of people unable to resist, and therefore the easier for the custodians of the empire to manipulate and wash the brains of those who are still considered as normal.

4.5.2 Phenomenal Brainwash

Without well-designed socio-political manipulations - charismatic speeches and promises - cleverly designed to hide actions usually undertaken in directions entirely opposite to that which the manipulative speeches and promises indicate, and without an expensive advertising policy to keep consumption-oriented desires of people at the highest possible level (and thus to keep people's ability to understand and unmask the manipulations, they are unceasingly bombarded with, at the lowest possible level), it would be difficult to support the global order.

The phenomenal brainwash in society serves to suppress the spontaneous expression of the self-organizing ability of social dynamics, that is, to prevent any organised network of people's resistance. If released, this ability could be a threat for the functioning of the empire, and for its highest pillars: the World Bank, the International Monetary Fund, the US Federal Reserve Bank and other incredibly rich organisations with silent and yet powerful influence on the global economy.

An army of police and security generously equipped with high tech means of surveillance constantly care for the smooth functioning of the global economy, that is, to ensure nothing impedes the growth of the profit of the economic bosses. The mass media, the politicians and the intelligentsia (those who belong to the latter are pressed to keep their mouths shut and support the bosses who pay them, or else easily lose their salaries and join the poor part of the society) do their best to inject fuzziness in the brains of the people. The most popular social slogans are:

- the governments care for all of us and do their best to make our life happier;
- the globalisation is inevitable and it provides you with abundance of things to buy and enjoy, so do not hesitate - go and buy what you want now!
- the banks and financial corporations are here to help the realisation of your dreams, so go and take another loan!

- the sustainable development, clean technologies and the cloning of genetic mutants will help you deal with the environmental problems, so do not worry about the pollution, buy another car!

- all of us have equal opportunities to realise our potential - to study, work, create, travel and live healthy and happy lives;

- the global financial institutions take care for the poor countries, so there are no reasons to worry about those who die in starvation and poverty, suppression and genocide;

- the wars that the strongest in the empire initiate (like the war against terrorism) are for the benefit of all, so support them, become a soldier, kill and die as a hero!

- the most efficient bombs are made to kill only people; if you have gold and money, they will be kept safely in your banks;

- if you feel tired, exhausted or depressed, try the new drug, lay down and feel blissful!

In one or another form, the content of the above slogans is repeated through the media, continually injecting delusive fuzziness or direct lies into the human minds. Behind each lie stand global economic interests, cleverly planned strategies aimed at more money, dominance and power.

4.5.3 False Identities

As far as the social dimensions of human individuality suggest its presentations outwards - to the other members of society - too much emphasis on these dimensions leads to an intense preoccupation with questions like 'what others think about me?', 'how do they judge my behaviour?', 'what could be their reactions to what I am going to say or do?', 'how to make others like me?', etc. This leads to the development of so called false identities, that is, individual identities designed to be shown in society.

The outward expressions of the false identity imply playing roles, carrying masks, speaking one thing and doing another, giving promises that are never fulfilled, expressing feelings that are insincere, telling lies. These actions are tools for a twofold fuzzification - we are deliberately fuzzifying something which is inherently fuzzy in us - our knowledge about ourselves, about the enigmatic nature of our authentic inner dynamics.

The more often one's false identity is used, the more significant its effects on one's life - such that the false identity eventually might become the only 'real' identity of its bearer. This is likely to occur, if the use of the false identity is associated with an intense expectation of success (money, prestige, pleasures) in society.

As we discussed in the previous chapter, the understanding of our inner nature, of our real identity is a difficult process; it needs efforts and perseverance to explore and reveal one's individuality and thus to make possible its authentic realization in life. It is easier to adapt to others' images, to borrowed ideas and opinions; the rule for doing this is simple: perform in such a way, that the people on whom you depend like your performance. This certainly implies not having grudge with the powerful ones but praising them, repeating and agreeing with what they say, and never worrying about the meaning of words like sincerity, honesty and truth.

With false identities and masks on the faces, the social life becomes a kind of performance, more likely a drama than a comedy, which one must enact. This performance appears fuzzy - irreal and delusive - both to the individuals who perform and to the spectators, as one can never have clear understanding of the possible consequences of this performance.

We perform in schools, in families, at workplaces, in community - everywhere. As a matter of fact, society does not teach us how to live authentically but how to perform. So absorbed in playing roles and carrying masks can we become that when alone we are confused as to what role to play.

By encouraging us to perform, to play roles and carry masks, society makes us unable to see the social reality as it is. With false identities we see only false and hence distorted pictures. Therefore we are not a threat to the economic rulers; they are afraid of people with ability to perceive the truth - to understand their own true identities, the identities of the others and the nature of the imposed economic order in society. The authentic selves are dangerous - they strive towards truth and wisdom, they do not tolerate social unjustness and corruption, they do not want to be deceived and exploited, they resist.

4.5.4 Towards Establishment of Global Control

The most important function of the global society is to exercise control. The economic elite must have information, as full as possible, about people who dare not obey the system; the latter will be classified as potential terrorists, as no democratic ways to change the essence of the established system exist any more. They are blocked by the global economic (and therefore political and military) power; the terrorist attacks, however inhuman their consequences, seem to be the only possible way to express people's resistance against the establishment and shatter its security.

If you decide to protest in the countries with the Western type of democracy, you must beforehand inform the authorities about your intention to do this, ask for their permission and, if it is given, perform your 'protest' strictly following the prescribed rules, under the close surveillance of the police. It is hard to believe that such kind of protest can affect the Global Establishment.

The prevention against any potential disorder in the empire needs well-organised global control of the human identities. Thank god, there are supra-computers which can help to realise the sweetest dream of the former KGB - a total control of human identities.

4.5.4.1 System of Fixed Identities

By educating and providing us with social, political, economic, national, religious, cultural, etc. milieus, society offers an open list of socially constructed identities for us to choose and learn how to fit, adjust and adapt to what they describe and require. We are 'invited' to identify ourselves with the identity of a student, teacher, politician, priest, pensioner, unemployed, poor, rich, etc. Our choices depend on our economic status, gender, nationality, culture, education.

As the guardians of the establishment need efficient - systematically organized and reliable -indicators of people's identities, they put huge amount of money to have these indicators aggregated into social groups and categories, to push each group into a special 'box', to label the boxes with numbers, codes and names, and to take care for their constant actualisation.

The establishment of a computer-based system of fixed identities (SyFI) helps authorities to exercise easy their control functions. It is much easier to supervise a System of perfectly organized files of data scrupulously collected to specify as full as possible selected aspects of people's identities than to deal with a continuum of individualities which are unstable, elusive, flexible, resistant to control and able to change and unfold.

SyFI operates with different tools - tax file numbers, licenses, passports, identity cards, certificates, diplomas, permits, visas, passes, finger prints, medical and genetic descriptions, etc. Each list of fixed identity characteristics is converted into a string of alphanumeric characters compressed to fit any computer database of any purpose. Once captured in the computer memory, human identity becomes imprisoned for life in a monstrous System from which no escape exists. And this is a great relief for any surveillance and bureaucratic institutions in society (there are millions of them!); when classified and labelled, we are predictable, it is easy to deal with us - to control our behaviour, to deceive and exploit, to wash our brains.

4.5.4.2 Control of Human Identities

SyFI resembles a procrustean bed. The name of Procrustes is well known from the ancient Greek legends as a name of brigand who used to stretch or mutilate his victims in order to make them conform to the length of his bed.

The custodians of SyFI in the Western democracies act in a similar way; the system is designed to operate most effectively with the two largest groups of human identities - the group of those who are actually poor - an ever-increasing army of unemployed or casually employed people, and the group of those who are potentially poor - the majority of so called 'middle class' trapped in lifelong payment of constantly growing mortgages. Everyone belonging to these two groups is 'invited' to fit the 'bed' labelled with his or her identity. If somebody resists, there are forces strong enough to handle any disobedience; beyond these forces stands the silent minority representing the most powerful (rich and influential) individuals and organizations in society, whose exclusive power to decide and act guarantees them freedom from the tentacles of SyFI. (According to many independent researchers, the owners and managers of the private Federal Reserve Bank of the US belong to such a rich minority; the gigantic capital of this bank allow its governing board to influence with their decisions and actions the world economy in ways, which make them always win.)

With identities imprisoned in a System ruled by forces over which the influence of the ordinary people is zero, we are turned into a herd of animals that cannot but obey The System. The information kept in the supper computers of SyFI is with incomparably higher controlling efficiency than the primitive cabinet files of KGB developed in the former communist regimes.

Advanced technologies provide SyFI with capacity to see, hear, read, and even sense what people are doing in each moment of their social and private life. In the 'free' world of the Western democracies, these technologies are permanently mastered under the generous support and control of the actual rulers of these 'democracies' - not the political marionettes for whom we vote during the election performances (all the marionettes are the same, no matter what is their party, so it doesn't matter for whom we vote!), but those whom the marionettes obey without objection (if they want to survive longer on the political stage and enjoy the generous access to the money of the taxpayers). The more efficient the control over people's identities, the safer and easier the life of the 'godfathers' of SyFI.

4.5.4.3 Forces of Separation

Everything that separates people from one another (egoism, greediness, possession, reputation, competition, conflicts, wars, hatred, envy, jealousy), no matter at what scale of manifestation - between individuals, between organizations or between nations - acts as a catalyser strengthening the isolation between the identities, making their boundaries hard, and thus facilitating the functions of SyFI.

While acting towards separation, we tend to:

- Intensify our competitive zeal, intolerance and constant fighting for acquisition of personal power, money and prestige, which inflates our individual egos with all their sick fantasies of success and dominance over others;

- Look at the world through the eyes of what can we grab from it in order to satisfy as much as possible the ever-emergent egoistic desires and ambitions;

- ➢ Collaborate with the supporters of SyFI in their efforts to keep our identities locked into boxes and cases of different nature and thus to make difficult the process of communication between us.

Ever growing is the abyss in communication between the identities of countries labelled as "rich" or "developed" and countries called "poor" or "developing". We are witnessing many cases of explosions of racism (abyss in communication between different racial identities) and suppression of indigenous cultures (abyss in communication between different cultural identities) in many places of the world. As we already mentioned above, with every day the abyss between the rich and the poor people in society becomes deeper and wider.

4.6 Education and Learning in Society

In today's society, the processes of education and learning are predominantly towards acquisition of various kinds of expert knowledge - a knowledge which can be used for solving specific problems. This kind of learning is centred in our minds (conceptual knowledge) and bodies (practical skill), and crucially depends on the development of our ability to think in a rational way, to analyze and synthesize, to extract and study cause-and-effect relationships, to generate hypotheses and test them experimentally, to draw out logical conclusions and master skills for performing certain actions.

The major goal of problem-oriented learning is to reduce or eliminate fuzziness in the individual knowledge, so that we could find precise solution of a precisely defined problem. As we pointed many times in this book, such kind of precisely defined problems cannot be meaningfully formulated the context of human-life complexity, be it individual or social. And if one specially creates 'precisely defined problems' to reflect the unpredictable dynamics of social complexity, these solutions will hardly make any sense in the turbulence of one's life. If the 'experts' persist in imposing them on the life dynamics, the consequences could be disastrous. Hitler and Stalin wanted to impose their precise political and military solutions on the life dynamics of the human beings; the results were catastrophic. Similar tendencies to use simple formulations and solutions when dealing with social complexity, at the level of economy rather than at a political level, are seen in the approaches of the custodians of the global order today.

"*Life is not a problem to be solved but reality to be experienced*" - these words belong again to Kirkegaard. Infinite is the number of levels through which reality manifests - from the macro level of the whole universe to the micro level of a single quark. And all the levels project on human experience - this is what social fuzziology argues in its theoretical framework. Not only because everything relates to everything else in the impossible-to-separate web of existential dynamics, but also because it is through our experience that we can grasp the meaning of the manifestations of these dynamics and ride on their inexhaustible power. We are endowed with a limitless potential to sense - recognize and understand - the meaning of the events of our experience. In every creative act of realization of this potential, a certain level of reality opens some of its secrets to us, a veil of fuzziness falls down from our understanding.

Unfortunately, the system of education in society does not pay much attention towards teaching us how to listen to and understand the 'voice' of our experience. Often, this voice appears too subtle, too soft and too fuzzy in comparison with the loud, sharp and determinate voice of the rational scientific models when trying to explain precisely how the surrounding world works and how the humans can utilise this 'precise' knowledge for the purpose of control over nature and people, systems and machines.

As we already discussed, the precise models do not make much sense when used to describe holistic phenomena and processes like those in nature, life and society. When applied to a description of such phenomena, any partial model, be it deterministic or probabilistic, precise or based on fuzzy logic, leads to delusion, to false views on reality - views which can be used for manipulative purposes in society.

Social fuzziology sees the **s**-fractals of human nature - the soul and spirit - at the basis of any holistic insight into the all-embracing unity of the existential dynamics and their timeless manifestations through the innumerous embodiments of the triad: universe, individual and society.

4.6.1 Metabolism of 'Borrowed' Knowledge

The educational institutions teach learners how to metabolise the so-called 'borrowed' knowledge, that is, a knowledge borrowed from outside the learners' experience and prepared by socially recognized 'gurus' and 'experts'. After pouring the borrowed knowledge into their brains, learners are asked to use it for purposes located again outside them, for example, for producing nuclear bombs and rockets, intelligent robots and self-organizing machines (who think and work obediently, unlike humans able to engage in unpredictable behaviour), genetic mutants (with totally distorted natural ability for realization of their inherent potential), cosmic stations (mainly for military purposes, espionage and global surveillance), and so on. Research institutions from all over the world fervently try to increase this kind of 'applied' knowledge; the most significant part of which is directed to serve their sponsors - powerful financial and industrial corporations - and to help them multiply their profits and satisfy their growing appetite for global power.

In the Western democracies, the biggest investors of the scientific research are the military-industrial complexes and the global corporations; the first thing they are interested in is ways to increase their economic power. Consequently, they endorse the individual efforts in this direction.

Sunshine is in abundance in the world but we continue to use petrol, pollute the environment and kill ourselves, because the huge chemical giants and the global car producers are not interested to support the research how to create efficient ways for using the solar energy. Research into quantum mechanics and plasma, computers and robots, cosmic exploration and genetic manipulations has strong military dimensions and due to this, the flow of money continues to go there.

The design, implementation, development and continuous innovation of the human-made systems require expert-type of knowledge and because of this, the educational system is forced to teach learners how to accumulate and apply such knowledge.

The more we immerse in the expert knowledge and its requirements for certainty and precision, the narrower the niche for researching ourselves becomes, the less able we are to hear and understand the holistic voice of our own nature, to interpret the fuzziness inherent in the events of our conscious and subconscious experience. Of course, society does not care about this; society wants computer experts, designers of robots and cosmic ships, genetic engineers and military constructors.

Who cares about the development of our intuition, of our ability to connect the events of our experience with the unfolding of the universal dynamics, or to see the occurrence of future phenomena in the immediacy of our inner dynamics? Who cares whether the existence has or has not a timeless essence, whether the human dynamics annihilate in the moment of death or continue in a state of potentiality? Society is not interested whether the individuals grow in wisdom or waste the most precious years of their life in serving the stupidity of the social commanders and their insatiable will to power.

It is the individual alone who is responsible for his or her growth, once awakened from the state of ignorance, in which society holds, unconsciously or deliberately, its members

4.6.2 Learning in Captivity of the Global Order

We are pressed to learn how to fit and serve the order of the global capitalist system - order based on unequal distribution of economic power and therefore ruled by those who have such power in abundance. Education is in captivity of this system, has its financial support and therefore obeys its orders.

Since their first days in school the children are taught how to contribute in making the system stronger. They are not asked maieutic questions, not explained that together with the external world around them, there is another world - much richer, much broader and deeper - the world of their inner experience. They are not taught how to be aware of this experience, to explore and understand it. Nobody will show them how to awake the creative power of their inner dynamics, how to rely upon it and minimise their dependence on the external 'help', how to control this power and use it not only for their own self-realisation and growth, but also for helping others create and grow.

The establishment of global capitalist economy is incompatible with people who know how to grow in consciousness, how to go beyond the limitations of the fuzziness in their understanding and reveal the delusion which the promulgators of the global order spread. It is much easier for the bosses of the system to manipulate a herd of narrow-minded 'experts' and precise 'specialists' than enlightened human beings with broad understandings of social reality; the latter can be a serious threat for the establishment.

As long as the process of education is under the surveillance of the global capitalist system, it is nothing but a scientifically informed brainwash, which instead of emphasising the personal growth of the learners, teaches them how to better fit into the system's requirements, to follow its rules, and to remain mesmerised by meaningless dreams for consumption-centred happiness.

4.7 The Health Illusion

In the vortical structure of the Vedic model of the inner human dynamics discussed in the previous chapter, the self-organizing forces that emerge out of each vortex are responsible also for the self-healing capacity of the individual. The locations of each of the Vedic vortex coincide with the location of nerve ganglia, glands and plexuses in the human body, which are considered by the health practitioners as vital autonomic regulatory centres; for example, there are vortices centred in the thyroid, hypothalamus, solar plexus, pulmonary and cardiac plexuses, reproduction organs.

As the vortices represent whirling energy patterns emerging out of the inner human dynamics, they cannot be borrowed from others or implanted from outside of our inner nature. No doctor in the world, no matter how competent s/he is, can keep them moving; the individuals alone are responsible for their functioning. In order to understand this functioning and to support it wisely we need the help of our own consciousness, of our experience, of our inner impetus to live and know.

Through studying how to concentrate and relax our mind and bodies, through practicing techniques that help us acquire inner peace and harmony, the flow of energy coming from the natural environment can be consciously directed inward and used to activate the vortices and therefore our self-healing capacity. Otherwise, the latter remains in a dormant state, and we need to rely entirely upon help from outside, when feeling sick. By doing this, we substitute the holistic effect of realisation of our self-healing potential with short-term partial effects produced by the use of various chemical drugs - medicines.

The more intensively we use medicines (and one cannot help but keep using them, as the effect of each dose is only temporal), the more addicted we become and the stronger the numbing effect that the 'curative' chemical substances exert on our self-healing potential. Eventually, this addiction results in losing our self-healing capacity.

So many people in the world die as victims of today's great health illusion that the help for our health comes from outside us! The society continues to amplify this delusion, because strong economic forces are behind it. The global pharmaceutic corporations make an unbelievable amount of money on this delusion; a great number of medical practitioners keep this delusion ever so powerful.

There is a socio-ecological explanation of the ever-increasing massive use of medicines in today's society. The more polluted the natural environment, that is, the more saturated with health-detrimental chemicals, the less efficient the realisation of our individual self-healing capacity, as the latter crucially depends on the support of such basic natural resources like air, water, sunshine, plantation, etc.

When

- the air is full of carbon dioxides produced by our cars and the industrial complexes spread all over the world;
- the dangerous chemical wastes, including nuclear, continue to be released in monstrous amounts;
- the soil and the water are irreversibly contaminated and the rains are acid;
- the process of deforestation and extinguishing natural species goes with an ever-increasing tempo,
- the ozone holes make the sunshine spread cancer in our bodies instead of healing them;
- the fruits and vegies are full of chemicals or 'genetically engineered' to look commercially attractive but detrimental for our health,

the efficiency of our self-healing capacity goes down, and we look for the use of medicines to help us.

So we are entrained in a kind of health-damaging vicious circle:

We continue to pollute nature with one kind of chemicals and at the same time fight the effects of this pollution on us by using another kind of chemicals. The more we pollute nature with the first (technological) kind of chemicals and thus gradually convert it into a source of new emergent illnesses, the more we use the second (medical) kind of chemicals to fight the illnesses and thus become gradually addicted. In the both cases, the result is one the same: serious destruction of our health.

Is there any way to go out of this vicious circle? Yes, if we take care about the natural environment and help it restore its own self-healing capacity; this will facilitate the increase our self-healing capacity (as we are 'children' of nature and our health totally relies upon its support!) and help ourselves reduce our dependence on medicines.

One essential aspect of the multi-faced mission of the social fuzziology is to show the fatal danger of the reliance on the help of medicines, while neglecting the vital factor for our health - our capacity for self-healing. The Mother Nature has endowed us with this capacity at the moment when we emerged out of her womb, and it is a grievous failure not to develop and use it to the full. Nature is the main supporter of the self-healing capacity. It is her generous and free supply of energy - her sun and air, water and soil, flora and fauna, harmony and beauty - that help the vortices of the human health move and generate their healing forces.

Self-healing is a holistic phenomenon - an expression of the self-organising ability of the individual as a whole, and there is only one way to stimulate it - through holistic means. Such are the means of nature! Thousands of years ago, this was fully understood by the creators of *Ayurveda* - the ancient Indian system of health (*ayur* means life and *veda* means knowledge in Sanskrit), according to which no single agent by itself can brings health. Ayurveda views the person as a composite of the same primary forces - air (force of expansion), water (force of adhesion) and fire (force of transformation) - which compose nature as well. When these forces act harmoniously in the individual, that is, in the way as they act in nature, they fulfil three functions: digestion (generating inner energies), absorption (sustaining the inner energies) and elimination (release of worked-off energies). These functions when considered holistically - in their simultaneously physical, emotional, mental and spiritual realisations - create health. Ayurveda defines health as soundness of three inseparable wholes: body (*shrira*), mind (*manas*) and soul (*atman*) (John, 2001).

How similar is this understanding to the harmonious resonance described in fuzziology, where the body-mind-soul resonance makes it possible to activate the bootstrapping algorithm of fuzziness and thus to transcend its limits in our understanding! The same resonance is seen in the ancient Ayurveda as activator of the vital 'bootstrapping algorithm' of the individual self-healing - a process of transcending the fuzziness of our knowledge in dealing with the human illnesses with the help of our own inner dynamics.

The earlier in life we understand the wisdom of the ancients regarding the vital role of nature in the conscious developing and strengthening of our self-healing capacity (which is in abundance when the organism is young and full of vigour), the more efficient the realisation of this capacity.

So, another aspect of the mission of the social fuzziology relates to the health education of the young people; this kind of education is a key factor in the promotion of health in society.

4.8 Dealing with Hard Negotiation

As the sources of 'borrowed' knowledge, that is, knowledge offered by books, experts, school teachers, university lecturers, mass media, world wide web, etc., are common for the whole society, people who have access to these sources share similar fuzziness in understanding the world and society. This is the fuzziness transferred into the human minds through the borrowed knowledge.

Although social fuzziology is critical to the massive use of borrowed knowledge in today's era of intense globalisation and phenomenal brainwash, there is a sphere of social interactions where the fuzziness induced in the human minds by the borrowed knowledge is practically useful; this is the sphere of human communication. While bombarded by similar flows of explanations, clarifications, comments, etc., people are more likely to share more or less similar viewpoints. This helps them when seeking consensus or negotiating certain issues of mutual interest.

Another factor helping the negotiation process is the similarity in the characteristics of the human nature. As we mentioned in the beginning of this chapter, our ability to recognise and react in a certain way to a certain feature in the character in another person is, up to a high degree, because we already have this feature in ourselves.

Social Fuzziology considers the similarity in the manifestation of the individual features helpful in the process of negotiation, particularly if the issue of negotiation is difficult or there are serious obstacle on the way to reaching an agreement between the negotiating sides.

4.8.1 Four Typical 'Hard' Situations

Wide is the spectrum of negotiations in society: from dialogues within the self or between close allies, lovers, spouses, friends, to struggles within organizations or across classes, cultures or groups, or between nation states on the world stage. Failure and breakdown of negotiations is also endemic, leading to untold suffering, in broken marriages, ruined friendships, strikes and lock-outs, terrorism and open or covert war.

A negotiation process is classified as hard if the participating sides are unaware how to facilitate its unfolding (either because of the complex and/or unknown nature of the issues under consideration or because of other sources of complexity and uncertainty), so as to make its outcomes satisfactory for each side. The following situations are typical of a hard negotiation process:

- Sides stick to entirely opposite views without willingness to make concessions or compromise;
- Sides do not know how to get out of a suddenly emerged deadlock in negotiation, even if there is a mutual willingness to move further;
- Some of the participating sides are interested in keeping the negotiation process in a state of deadlock for a certain period of time and therefore resist any further progress;
- There are powerful third-party forces that impede negotiation, despite the willingness of the participating sides to reach satisfactory outcomes.

Situation 1. Polarisation

The attraction to mutually opposite positions ('polarities') strongly energizes the negotiation process and leads it to a critical state - a state in which human interactive dynamics become critical (turbulent, chaotic).

Criticality of the human dynamics resembles the self-organised criticality in nature - a wide spread complex phenomenon studied in detail be by Bak (Bak,

1996): complex dynamic processes in nature tend to reach critical states impregnated with potential for spontaneous emergence of ‘avalanches of changes'. In the critical states, minor perturbations can trigger significant changes in the future unfolding of the processes under study.

In the context of the hard-negotiation process, Back's theory has an obvious interpretation: if the human interactive dynamics expressed in the negotiation process reach the critical state, it would be enough for one of the participating sides to slightly move from an initially chosen position (viewpoint, stance, opinion, belief, attitude) and the process will give birth to unforeseen opportunities for further consideration by the negotiation sides.

Sometimes, where the negotiation process has yet not reached its critical state, it is possible to apply a 'softer' approach: to help the sides, through a skilful facilitation, become aware that their positions, initially appearing as diametrically opposite, have no hard and insurmountable boundaries. By slightly changing (expanding, fuzzifying) each position, the whole process might 'move' in a mutually satisfactory direction.

Situation 2. Emergent Deadlock

If the negotiation process is stuck at an emergent deadlock and the sides do not know how to overcome it, the approach of fuzziology can be applied in the following way: the facilitator may try to initiate the effect of resonating fuzziness, that is, of a spontaneous overlap of fuzzily expressed (open for sharing with each other, flexible) thoughts of the participants about how to proceed further. There is alway fuzziness in human understanding: what appears 'absolutely clear' for some of the negotiating sides (and used by this side as an obstacle for further development of the negotiation process) can be turned into a densely fuzzy argument, if some additional (or new) aspects are taken into consideration. There exists a version of Peter's law in fuzziology: *for any well-defined point of view (opinion, argument, standpoint, position), one can always find an equally well-defined and entirely opposite point of view*. However unserious this 'law' might sound, it acknowledges multi-dimensionality of any process of negotiation, no matter how simple or hard this process is.

It is much easier to facilitate a resonance on an issue about which people honestly admit their fuzziness than pretending that their knowledge is exact. When the expressed views are treated as 'exact', the agreement can be reached only if these views coincide; the chance for such coincidence is very law in complexity of the hard negotiation. One of the reasons for lack of success in the negotiation between the Israeli and Palestinians is that each side knows 'exactly' that the other side must do in order to stop killing each other.

The effect of resonating fuzziness stimulates emergence of creative insights in the negotiating process.

Situation 3. Imposed Deadlock

Where each of the sides has a high degree of willingness to reach a mutually satisfactory outcome, there is a higher probability that the resonating fuzziness of their conceptual positions will result in transcendence of the negotiation deadlock. But if one of the participating sides has some interest in keeping the negotiation process in a state of deadlock and thus creates obstacles to any further development of the process, social fuzziology has the different task of dealing with motives behind such behaviour.

In such situations, motivation is a fuzzy category which always reflects a wide and complex spectrum of human thoughts, feelings, expectations, beliefs, hopes, endeavours, etc. as they emerge or disappear, persist or evolve. However strong a motivation might be, its nature can never be described in a black-and-white way. The inherent fuzziness of motivation makes its boundaries permeable, that is, open to influence and change. It is this openness which matters most in negotiation.

By exploring in depth the motivation of all participants, it is possible to facilitate conditions in favour of emergent changes in motivation. As far as any imposed deadlock pushes the negotiation process towards criticality, the 'butterfly effect' releases its 'moving' power; even the tiniest changes in the motivation of the side responsible for an emergent deadlock can create significant changes in the overall development of the negotiation, thus moving it eventually to satisfactory outcomes.

Situation 4. Third-party Influence

Third-party forces can affect the negotiation process in two ways:

(a) during the negotiation phase: through participants who are supported by these forces;

(b) during implementation phase: by impeding the implementation of the achieved negotiated outcomes.

In case (a), if a participant supported by an external force tries to impose deadlock on negotiation, a strategy similar to one considered in Situation 3 (Imposed Deadlock) can be applied. The difference is only in the location of the source of motivation; most probably the source of motivation for impeding the negotiation process is in the external force rather than in the participant supported by this force. No matter where the source for motivation is, the sides need to explore the fuzzy nature of the motivation and to facilitate emergence of conditions (be they at local, national or international level) in favour of motivational shifts. The rest is again in the hands of the butterfly effect; when negotiation enters the critical zone, even a tiny motivational shift becomes crucial for negotiation to escape from the deadlock.

Case (b) is more difficult as it depends on the distribution of power in society. If any particular negotiation outcome meets the resistance of a force with significant power, the realization of this outcome is under threat.

The approach of social fuzziology can help in the following way. Because any social issue has many dimensions, and any agreed outcome of negotiation inevitably relates to a certain social issue, it is possible to elicit those dimensions of the outcome which will not evoke, at least initially, any fatal blocking resistance from the force with significant influential power. Under this condition the process of implementation can commence. Thereafter it is the breadth of the overall dynamics of social complexity that becomes decisive for the unfolding of the process. So the rule of thumb is be well prepared and proceed to take the first step; be aware of the consequential changes and use this awareness as a guide when taking subsequent steps.

4.8.2 Pursuit of Consensus: Practical Aspects

The approach of social fuzziology can be effectively applied to any process of consensus seeking. The awareness of the fuzziness embedded in what the sides consider as known in respect to the hard negotiation process in which they have been involved, makes them more flexible when discussing each side's position and thus more prepared to act together.

The 'algorithm' of applying social fuzziology in consensus seeking is very simple: instead of considering one's own opinion (position, viewpoint, stance) categorical, determined and final, one looks at it as a product of knowledge which is inherently fuzzy, open to changes and non-final, and therefore without hard-to-surpass boundaries.

The realization in practice of this 'algorithm' by the sides participating in a hard negotiation process facilitates their search for mutual understanding, concessions and compromises, and thus helps them eventually reach consensus.

Although the above approach sounds simple and transparent, its realization in practice is not easy as a result of three major factors:

- To accept that one's knowledge is fuzzy means to be humble - a quality that is rarely displayed in today's competitive and power-based society.

- To accept that one's knowledge is fuzzy means to be ready to move beyond any dogmatic viewpoint extracted from 'borrowed knowledge', that is, knowledge from books, 'gurus' and 'experts', and to directly perceive the negotiation process as it is. Unfortunately, throughout life we have been taught how to apply, and rely upon various sources of borrowed knowledge. Hence the intuitive component is not well enough developed to guide us when we are seeking consensus.

- To accept that one's knowledge is fuzzy means to be able to go deeper into one's own experience and reveal paradoxes and conundrums that nobody else could have perceived, as they belong to one's own way of experiencing life. To work in this way is not an easy task. It needs a great amount of alertness, sensitivity and thirst for knowing oneself, as well as the surrounding world.

Although the points above describe quite serious obstacles on the way to consensus in hard negotiation, no one of them is fatal to the continuation of the negotiation process. The mere awareness of the existence of obstacles, combined with honest endeavour to understand the forces behind them, acts as a stimulus towards seeking consensus. The higher the willingness of the participating sides to stay in an open dialogue, the more prepared they are for joint action towards consensus.

4.8.3 Transcending Paradoxical Situations

Hard negotiations often have roots in intricate paradoxical situations, which cannot be resolved using rational thinking. Logic, be it black-and-white; fuzzy; inductive; abductive; or deductive, is helpless with paradoxes. The way to deal with paradoxes is to transcend them, that is, to bring forth conditions under which the paradoxes dissolve. This is the way of 'bootstrapping', the way of facilitating the emergence of consciousness resonance - the way of social fuzziology. We succeed in transcending the fuzziness inherent in our knowing not by trying to reduce or eliminate it, but by linking (connecting, overlapping, superimposing) all available pieces of fuzzy knowing and energising them with our motivation (thirst to know, impetus for self-realisation of our creativity, yearning to understand and make the power of our spirit work) so as to allow the emergence of a 'higher level' type of knowing.

For example, the fuzziness accumulated when trying to explain the phenomenon of relativity using the tools of the Newtonian paradigm brought about the new paradigm of Einstein as a 'higher level' of knowing, in which most of the fuzziness related to the paradoxes and discrepancies in the Newtonian paradigm were dissolved. The new paradigm appeared much more adequate to the observed phenomena and much more clear in describing them. Subsequently, with the further development of this new paradigm, other paradoxes and discrepancies emerge as new sources of fuzziness, which persistently 'fuzzify' the knowing related to the paradigm of Einstein. This is the natural way of expanding and evolving of our consciousness.

4.8.4 The Incoherence Principle

The famous paradox of K. Arrow, known as Arrow Impossibility Theorem (Arrow, 1963), reveals the impossibility of generating a socially satisfactory choice from deterministic (strictly defined) choices of individuals making up a society. In other words, it does not matter what kind of group choice rules are applied, they cannot help in aggregating the deterministic viewpoints (positions, opinions, preferences) of the participants into a deterministic viewpoint of the whole group, which is compatible with (acknowledges, takes into consideration) each participant's choice. According to Arrow's Impossibility Theorem, the group choice will be dictatorial - imposed either by some participant in the group or from outside the group.

The paradox of Arrow dissolves if the individuals' choices are non-deterministic (Dimitrov, 1976a). According to the so-called Incoherence Principle (Dimitrov, 1983), the more certain (non-fuzzy, determinate) each individual's choice, the more uncertain (fuzzy, indeterminate) the choice of the whole group, and on the contrary: the more fuzzy the individual choices, the greater the chance for the group to negotiate a socially satisfactory deterministic (categorical, non-fuzzy) choice.

When applied to negotiation, the Incoherence Principle becomes a tool to achieve agreed outcomes (actions), which take into consideration and fully acknowledge the fuzzy choices of each side participating in the negotiation. As far as the approach of social fuzziology is rooted in the fuzzy individual choices, it provides an efficient leverage point to transcend the paradox of Arrow in the process of hard negotiation.

Logic would require a socially satisfactory correspondence between deterministic individual choices and the deterministic group choice, that is the deterministic choice of each individual to be mapped into a deterministic choice of the group, which satisfies all the individuals. But the paradox of Arrow denies the existence of such mapping. The Incoherence Principle offers a kind of anti-logical way: it is the fuzzy individual choices that can lead to a deterministic and yet socially satisfactory choice of the group as a whole. This is consistent with the way social fuzziology looks at the negotiation process; by acknowledging that fuzziness is inherent in the human knowing, social fuzziology strongly supports the fuzzy expression of the individual choices.

If the participants involved in hard negotiation are committed to reach precisely formulated negotiation outcomes (actions), satisfactory for each side, their positions need to be formulated in a fuzzy way.

The above theorem epitomises the phenomenon of transcending fuzziness while using fuzziness as a leverage point: the fuzzy (indeterministic) positions of the sides participating in hard negotiation can result in precise (deterministic) outcomes - actions. But such a non-fuzzy transcendence does not last long. The ever-changing dynamics of social complexity are hardly compatible with long-lasting negotiated outcomes, particularly if their emergence is result of hard negotiation.

What needs to be strongly emphasised is that the finding of social satisfactory group decisions has very little to do with the rightness or truthfulness of these decisions. They are decisions which happen to satisfy the sides participating in negotiation; for the rest of the world that may appear wrong or oppressive or irrational or even stupid

As we pointed out in the section 4.8, the possibility for reaching an agreement depends on the degree of similarity in the fuzziness of understanding the issue of negotiation by the participating sides. This fuzziness could be so dense that the agreed decisions may appear nonsensical or wrong for people who have deeper and broader understanding not only about the issue of negotiation, but also about the conditions under which the negotiation process has been carried-on, about the level of competence of the negotiating sides, etc.

4.9 Conclusion

Although this chapter was critical about the establishment of today's global society and the obstacles which it creates on the way of the individual endeavours for self-finding, self-fulfilment and growth in understanding, it is clear that no one individual can live without society. We are social creatures and since the first day of our appearance on the earth we are destined to be together. And it is because of our togetherness, that we have developed ability to think and feel, speak and act, understand and know.

The individual insights enrich the accumulative knowledge of human society. And vice versa, the society continually teaches us its lessons-in-vivo - lessons of compassion, forgiveness, tolerance, selflessness, modesty, humbleness, mercy, courage, altruism, generosity. In this sense, society is a powerful facilitator and catalyst for changes in our nature.

Great and long-lasting are the effects of those lessons of society, which deeply move us and leave profound traces in our memories; often they relate to events that have made us suffer, felt hurt, neglected, exploited, insulted, abandon.

The experience of this kind of events makes us contemplate on the sources and consequences of our suffering, and re-think our relationships and behaviour. Gradually, an understanding might arise about the deepest hidden roots of the human suffering.

Powerful are the lessons of society on the ethics of our behaviour, which show us that it is hard to behave in a fuzzy way - partly good and partly bad, partly honest and partly liars, partly helping and partly killing. These lessons reveal to us that the slightest touch of egotism, cruelty, snobbery or vulgarity in the relationships can spoil everything done in the twinkling of an eye. And this is valid not only for the relationships between friends, but also between colleagues, relatives, and even between people whom we meet for a first time. The wisdom of Zen says: "A hair's breadth of difference, and they are Heaven and Earth apart" (Blyght, 1983).

In the same book on Zen, there is an interesting description of the rightness and wrongness of the majorities and minorities in the human society:

"Majority will always persecute minorities, but the majority, thought never right, is not always wrong (usually they are just nothing at all!) and therefore the minority, though never wrong, is not always right."

There is a deep meaning in this seemingly fuzzy interplay or words. We cannot blame only the society (on the basis that it is never right) for our failures and ignorance. Being rarely wrong to ourselves (as there is nobody in the world who know our own nature us good as we do), we are not always right; so, better to look for the sources of our not-always-rightness firstly in us and than in society.

The go-beyond approach of fuzziology, when applied to the fuzziness of our understanding of society can be formulated in the following way: "*Do not reject society, but do not accept it either! Go beyond its repulsive and attractive forces*!" What is the meaning of this fuzzy formula?

It is clear that we cannot reject society, we cannot live by constantly opposing or isolating ourselves from it; we need its lessons, we need people's love and help, people's warmth and compassion, encouragement and support.

But there are many things in society which we resist accepting; maybe the first among them is its suppressive power trying to turn the majority of us into creatures without voice, without conditions to realise the most precious gift that each of us has - our potential to create and evolve in consciousness and spirit.

And "*Go beyond*" means understanding society, initiating the 'bootstrapping algorithm' of the fuzziness that dwells in this understanding. Socrates - the patriarch of social fuzziology - has left the simple formula for us how to go beyond any limitations on the way towards self-fulfilment and wisdom: "KNOW THYSELF!"

5 Case Studies: Understanding Fuzzy Social Categories

5.1 What is 'Australian'?

In social communication, most of the words are fuzzy, and this is the key to their social use.

For instance, 'Australian' fuzzily denotes people or objects coming from Australia, and its connotations are much more diverse, different in different contexts. In USA they may imply 'Crocodile Dundee' connotations - simple, physical, easy-going, unsophisticated etc. Australia's distinctive fauna - kangaroos, koala bears - and its Aboriginal people are part of this set of associations. Its status as a former colony of Britain may also be part of the set. The Olympic Games may draw on and add other connotations - competitive, sporting, etc. These meanings are hard to pin down: they are fuzzy.

But it is in the fuzzy space of connotations that words do most of their work. The connotations of 'Australian' sold millions of dollars of tickets to the 'Crocodile Dundee' films, and supported a successful publicity campaign in America. They probably played a role in Australia winning the right to host the Sydney Olympics.

Australians have most of these associations, along with others that are more specific to Australia. This word carries the fuzzy meaning of an apparent unity which includes a diversity of sometimes conflicting elements and associations. For instance, for some it suggests the culture of the Anglo-Celtic majority (like 'Crocodile Dundee'), for others it includes Aboriginal people and culture and later migrant cultures.

'Australian' within Australia also connects with the full range of political and economic activities that Australians engage in without being uniquely or typically Australian: there is an Australian dollar among the world currencies, Australia forms military alliances, especially with USA in the Pacific area, and so on.

This field of associations includes many instances of contradictions: eg the unsophisticated 'Crocodile Dundee' and the futuristic Opera House, the military alliance with USA and the new role in Asia. The proliferation of contradictions in an open network of associations makes the connotative field not simply fuzzy but chaotic, that is, saturated with extremely context-sensitive and therefore unpredictable dynamic interpretations. We can represent this set of meanings of 'Australia' as follows:

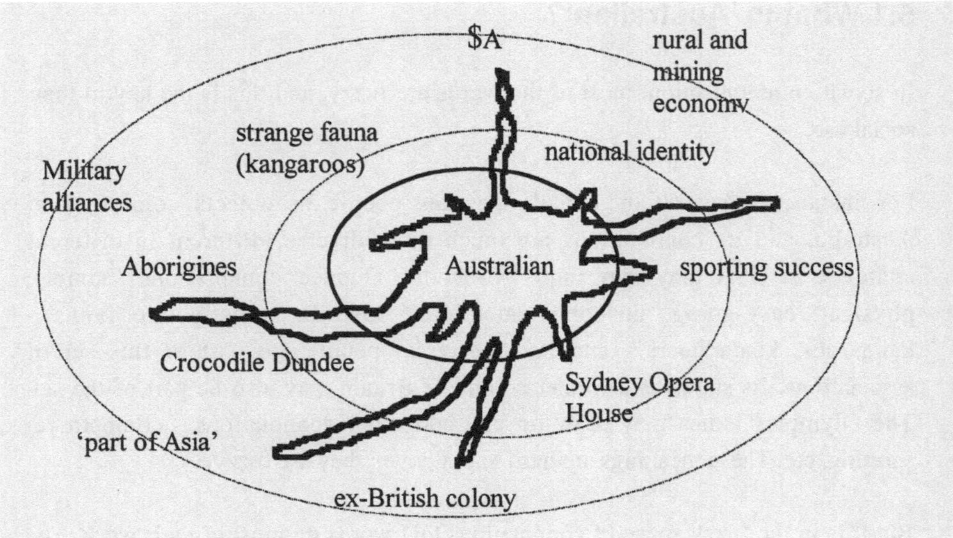

In this image, the central circle is the fuzzy denotative meaning of 'Australia', and the two outer circles represent ideas that are more or less strongly associated with 'Australia', its connotative field. But the circles are not crisp boundaries. They are arbitrary stages along a fuzzy continuum of strength of association. These links are represented as like axons of a neuron, with 'Australia' understood as being at the centre of a kind of neural network.

This is a simplified map of the two kinds of fuzziness around words: fuzzy denotative meanings, and the more chaotic connotative field.

Without some kinds of constraints meaning would quickly revert to chaos (uniform distribution of many possible interpretations, which eventually leads to meaninglessness), and there could be no stable inferences possible from any premises to any conclusion. It is necessary therefore to have a working idea of constraints in the analysis of the chaotic fields of associations or connotations.

The question 'What is Australian?' could have a single answer which is unitary, consistent and non-fuzzy, and that answer (or those answers, since no single proposal will be accepted by all or even a majority of groups), and that would be too tightly constrained. Yet there has to be a field within which the differences operate: a fuzzy field of proposed constraints.

5.2 Use and Abuse of Fuzziness in Social Life

There is a paradox in the role of fuzziness in social life, especially in dealing with such complexities as arise in the world under globalisation. On the one hand, fuzziness of thinking is indispensable for anyone seeking to understand this world better. Fuzzy knowing will not give certain truths for uncertain situations, but it will be a creative kind of response. This fuzziness can be called *reflexive*, since it reflects and resonates with the uncertainty of knowing and the known. However, often the level of fuzziness generated seems unmanageable to people, a hindrance to understanding and action. In this situation they become vulnerable to a widespread quality in modern life we called *delusive* fuzziness.

Delusive fuzziness differs from reflexive fuzziness in two respects. It always involves re-fuzzification from a prior act of defuzzification. That is, crisp knowledge is the only valid form for that thinker, and the re-fuzzification is designed not to make the message more complex and dynamic. It is always motivated to deceive and manipulate receivers, driven by the strong interests and crisp analyses of power in the relationship. This tactic makes use of the fuzziness imbedded in human understanding of social reality to misuse or abuse it.

Most speeches and promises of charismatic leaders in business and politics are full of delusive fuzziness, designed to evoke a high degree of fuzziness of people's understanding and hide motives and ideas very different from what the speakers imply or suggest. Delusive fuzziness not only obscures a complex picture of reality, it distorts our self- understanding.

To illustrate how delusive fuzziness works in political discourse we will take a speech made by President Bush on 22 September 2001, announcing to Congress and the nation the actions he intended to take in the wake of the attack on the World Trade Buildings and the Pentagon. This was a highly successful speech for its immediate constituency, the US people, generating a 91% approval rating for the President who not long before was languishing in the low 50s. How did it work, and what can social fuzziology say about it?

The speech began: "Tonight we are a country awakened to danger and called to defend freedom. Our grief has turned to anger and anger to resolution. Whether we bring our enemies to justice or bring justice to our enemies, justice will be done".

This speech pivots around a classic form of delusive fuzziness: the use of abstract concepts separated from contexts of application, so that they wait to be filled by speakers and hearers alike, in a creative act that is also wide open to manipulation. The two terms repeated in the speech were ‘freedom’ (9 times) and ‘justice’ (3 times). In delusive fuzziness, not only is the scope of the term extended arbitrarily so that it is not clear what it refers to (who is free? US citizens? Their politicians? Others in the world oppressed by American policies?). Nor is it clear how ‘freedom’ applies in this case. For instance, Bush claims the motive for the terrorists is their hatred of ‘freedom’; Americans are asking "Why do they hate us?" and he answers: "They hate our freedoms" (Sydney Morning Herald, September 22-23, p. 11). But Malaysian Prime Minister Mahathir, a Muslim leader, offered a different answer: ‘the anger among the oppressed.’ The point is not that this is a true and exact description of the motives of the terrorists, but that this is what they say is their motive, and Bush's statement is only successful with his American audience – as it undoubtedly was – because ‘freedom’ had first been emptied almost entirely of specific content, so that the extremely fuzzy residue could be more easily applied to the motives of ‘the enemy’.

The second tactic used in the speech is to use statements that initially seem crisp, clear and definite but which have an unmanageably large and unclear scope: "Every nation in every region now has a decision to make: Either you are with us or you are with the terrorists".

Its terms seem absolute and black-and-white, with no fuzzy hedges. It constructs the image of the President as resolute, decisive, making statements that are reassuringly crisp (reassuring for those who see the situation as too unclear). But the apparent crispness comes into conflict with other parts of the text. "Our enemy is a radical network of terrorists and every government that supports them". That is, the target itself is complex and diffused – the kind of fuzzy object thrown up by globalisation. He gives some figures: "thousands of these terrorists in more than 60 countries" – a large figure and highly unspecific. Is USA for instance among these countries? Is he declaring war on a third of the globe? If so, there is danger indeed, as he said at the beginning, but much of that is created by the interaction between the crispness of his choice, and the fuzziness of the world as he constructs it.

This speech, and the US policy it announces, is delusive, in that the President will have been advised by many experts how complex the situation really is when viewed on a global level. The terms he uses are both apparently 'simple', pairs of binary opposites, but as the public tries to make sense of them they are filled both with the fuzziness of different people's knowledge, desires and fears. It is also very successful, and typical of the rhetoric of many other leaders, in other administrations, in other countries.

The mass media clearly plays a crucial role, injecting fuzziness into human brains to go with an ever-accelerating tempo. Here is what one former chief of staff of the "New York Times," called by his peers "the dean of his profession", used to say to his colleagues: "I am paid weekly for keeping my honest opinions out of the paper I am connected with. Others of you are paid similar salaries for similar things, and any of you who would be so foolish as to write honest opinions would be out on the streets looking for another job. If I allowed my honest opinions to appear in one issue of my paper, before twenty-four hours my occupation would be gone. The business of the journalist is to destroy truth; To lie outright; To pervert; To vilify; To fawn at the feet of mammon, and to sell his country and his race for his daily bread. You know it and I know it ... We are the tools and vassals for rich men behind the scenes. We are the jumping jacks, they pull the strings and we dance. Our talents, our possibilities and or lives are all the property of other men. We are intellectual prostitutes" (this quotation is from a material distributed through Internet at http://www.teamlaw.org/).

This statement may sound over dramatic. Noam Chomsky, the distinguished linguist and critic of American imperialism, describes the process more dispassionately. His book "Manufacturing Consent : 'The Political Economy of the Mass Media" 'traces the routes by which money and power are able to filter out the news fit to print, marginalise dissent, and allow the government and dominant private interests to get their message across to the public... The raw material of news must pass through successive filters, leaving only the cleansed residue fit to print.' (Herman & Chomsky, 2002)

What Chomsky describes is a massive simplification – removing alternative voices and facts – which defuzzifies the news before adding the specious re-fuzzification of 'sensationalism'. He claims that these filters are not neutral attempts to adjust the complexity of the world to the simple capacities of the mass reader. Their negative role of filter is designed to allow the voice of government and powerful interests full sway. In the case of the 'war on terrorism', newspapers throughout USA as in Australia and other countries devoted massive coverage to the viewpoints of government spokespersons, and almost none to those who had a deeper knowledge of Islam in all its forms in a complex world.

Hardt and Negri (Hardt and Negri, 2000) deal with the crucial topic, of how 'global citizenship' can become a reality, with a majority able to participate with understanding in the debates facing the globe. (p. 398). We suggest that one part of the answer is to make available some of the basic insights and strategies of social fuzziology. The authors have in mind what they call 'the multitude'. We have in mind also all educated citizens, including scientists and engineers, who are typically kept entrapped by the information they derive from the media and other sources, and whose training does not alert them to the need to react differently.

This is because acquisition and making sense of social information crucially differs from the analogous processes in science and engineering. In the latter we collect information in order to reduce the fuzziness imbedded in our knowledge of nature (reflected in various branches of the natural science) or human-made environment (reflected in the engineering science and technology). The approach of reducing fuzziness does not work with social information because society is not separated from us. We are society - it consists of us, and we also constantly create, destroy and accumulate an infinite amount of social information by sharing our experiences, our thoughts and views, feelings and emotions, beliefs and dreams.

At the same time, society crucially affects us in the process of exchange of social information encapsulated in its numerous multimedia incarnations. What we hear, watch, read and write influence our experience, the ways we think, feel, believe, dream, aspire, and thus create or destroy our capacity to be global citizens.

5.3 Fuzzy Borders in the Globalising World

Globalisation is not a single tendency in a single linear direction but the unpredictable and constantly changing product of two broad, contradictory interacting tendencies. On the one hand there is an increase in flows and dissolution of previous boundaries leading to the creation of new patterns of connection and new entities acting on a larger scale of coherence. At the same time, various agencies have access to new crisp, 'linear' technologies which create the possibility of even greater levels of ever more precise control. Hardt and Negri paint a standard picture of the emerging situation under globalisation: "The primary factors of production and exchange – money, technology, people and goods – move with increasing ease across national boundaries; hence the nation-state has less and less power to regulate these flows and impose its authority over the economy. Even the most dominant nation-states should no longer be thought of as supreme and sovereign authorities, either outside or even within their own borders" (ibid.; p. xi).

This is only part of a complex picture requiring a fuzzy map. The texture of life under globalisation is paradoxically marked by grander, more fluid and functional forms emerging from the chaotic forces shaping social economic and cultural life, 'order out of chaos', and by the disorder flowing from the intensified attempts by many powerful agents to impose linear, crisp forms of thinking in order to achieve control. Globalisation should not be identified with one or the other, but with new conditions under which both act and interact.

We take a particular case to illustrate. On August 28, 2000, there was a riot in the Woomera Detention Centre in Australia, an isolated place on the edge of the desert in the central part of the continent, where illegal immigrants are detained, specifically a number of immigrants who arrived by boat from the Middle East and claimed refugee status. The riot was quelled with tear gas and water cannons; 32 guards received 'soft tissue injuries', and 25 people identified as 'ringleaders' were removed to Adelaide and detained. The Immigration Minister, Philip Ruddock, blamed 'well meaning advocates' outside the centre for inciting this unrest. He threatened to cut funding to the Refugee Council of Australia, a source of these 'well meaning advocates', and announced a tightening of security measures at the Centre.

This incident shows nation states do indeed have effective power within their own borders, and sufficient resources to impose a linear solution on those who oppose them, who are a small, weak minority. However, the exercise of control was not cost effective. The actions they took to control the situation mostly had the opposite effect, causing a tense situation to spiral out of control. They intended to appear strong and resolute, and only appeared brutal. An editorial in a major national newspaper, *The Australian*, of August 30, 2001 (p. 12) was critical of this policy, under the headline 'Woomera riot reflects poor refugee policy' and commented: "The reflex action of condemning the detainees and tightening security is useless without understanding how the system fuelled the uprising". (Later on, in January 2002, this anti-human policy of the Australian Government exploded into massive hunger strikes of hundreds of people, including small children, hold behind the barbed wires of the detention centres all over Australia).

Essentially the editorial critiqued the lack of fuzziness in the government's categories and systems of control, with boundaries that were too rigid and crisp – barbed wire around the detention centres (reminding the barbed fences of the Nazi's concentration camps in Europe during the Second World War) - an invisible wall around 'Fortress' Australia. More important, this rigid, crisp thinking which was causing such problems co-existed with the fuzzy thinking at the basis of 'normal' practices.

The editorial criticises the rigidity of the categories used to describe these particular people. It points out that the overall category 'illegal immigrants' covers all kinds of immigrants without visas, including genuine refugees. It notes some people 'take advantage of the refugee approval process': that is, the category of 'genuine refugee' is fuzzy and some refugees can exploit the fuzziness. But it adds Australia has a policy of accepting 'genuine' refugees.

The Government in contrast, they say, has tried to have a 'refugee policy which is clear, and everyone who arrives illegally will be considered to have broken the law and be treated accordingly'. That is, it has adopted a control strategy that avoids all ambiguity and can therefore be applied simply, so that the laws are easy to define, but not just or appropriate. This suggests a corollary to Zadeh's principle of incompatibility: in complex situations, it is not possible for laws to be both precise and just.

In practice, as the Editorial points out, the Government does not only apply a single, simple. "Yet this Government is just as happy to treat those who enter Australia on tourist or business visas, and then deliberately overstay that visa, relatively leniently". This category of illegal immigrant is larger. Normally they arrive by air, not sea, as did the Refugees in this case. "Yet this second group has become the focus of hysteria over illegal aliens".

The Government policy has a fuzzy and a crisp interpretation of the same set of laws. The 'fuzzy' laws apply to the majority of cases, the 'crisp' to a minority. Yet it is the minority constrained by 'crisp concepts' who arouse hysteria in the public mind, leading to riots in detention centres which in turn fuel that hysteria. The 'rationality' of non-fuzzy rules in complex situations provokes irrationality.

From this analysis we derive that in the contemporary era, national governments in developed countries use both crisp and fuzzy categories to exercise control in the increasingly complex situations produced by globalisation.

Fuzzy categories and controls normally reduce conflict and have become part of the normal mode of governance ('normal' rules for the majority of people in most situations), while the crisp categories and controls normally exacerbate conflict, increase injustice and inhumanity, and destabilise the harmony of a group.

Although crisp forms of control are normally dysfunctional, they can serve the interests of particular governing groups, especially right wing or populist governments, in whom they signify strength though they produce weakness. Excessive and inappropriate crispness in critical situations provokes hysteria ('chaotic fuzziness') in some of a populace, from which governments or other groups hope to benefit.

5.4 'Mexican Americans', 'Chicanos' and Santana

Different commentators have widely divergent views on the fate of two crucial categories of social analysis in the wake of globalisation: culture and identity. Some see the tidal wave of globalisation as removing all individual cultures and identities, subsuming them into a single homogeneous transnational culture, a single global identity replicated across the planet not as a present fact but as an inevitable outcome. Others note that the opposite tendency also exists: the development of hybrid multicultural forms within or against both national and transnational cultures and hybrid identities.

From the point of view of social fuzziology it is not necessary or productive to choose between these two, or to choose at all. Cultures and identities in the global world are likely to show new connections, new unities, and new differences, as the seemingly more stable forms of earlier phases of history change and transmute at the edge of chaos. At the same time, aggregations of power increase in size and in the resources at their command, able to attempt to control others on a scale never dreamt of before. Both possibilities are part of globalisation, and they have been realized many times over.

What is happening today and what is likely or possible to happen tomorrow needs to be looked at by mobilising the fuzziness of thought. Under the pressures of globalisation, as many have observed, there are already major movements of bodies, goods, money, culture and ideas across borders with consequences that can be scanned for their virtual meaning. For instance, such a flow is happening between Mexico and USA at an ever-accelerating pace. Its consequences for new identities are already complex and diverse. On the one hand the US 'system of fixed identities' now contains the category of 'Mexican American', which is both a hybrid category and also a fuzzy continuum from with various grades of measurement, reflected in types of visa. From the US point of view it is intended as a 'hybrid' in the sense that it will be sterile – the offspring will be fully American in attitudes, ideology etc.

At the same time there is a large group of people who are outside the system of identity called 'illegals' by the US but referred to as *nodocumentados* by the people concerned, taking pride in the fact that they are outside the system. But many businesses rely on the cheap labour of these people, in factories or in home duties, the cheapness coming from a fuzzy status which is different and more dynamic than contained in the fuzziness of 'Mexican American'.

Alongside the official identity system there is another one that takes some of the identifiers – in this case 'Mexicanness' – and fill them with rich meaning. This has happened with people who identify themselves as 'Chicanos' (a corruption of the Nahua-Spanish word pronounced Me-hicano). This word is not crisp but full of fuzzy meanings, comprehending a sense of shared values, shared history, shared identity. In crisp terms it may seem to correspond to 'Mexican American' but it does not do so in normal practice. Many people would accept an identity as 'Mexican American' but deny they were Chicanos. 'Chicano' has become the focus of what is called 'identity politics'. It carries greater social force than any crisp term for social identification, and this in fact is a general rule. Qualities abstracted from the fuzzy richness of social life to be used in a computerised system of information or control will never carry layers of social meaning needed to understand a people or a culture.

A term like 'Chicano' is a rich bearer of identity, but it is still built around a reductive binary, 'us and them'. It is fuzzy, but only to a degree. Even greater levels of fuzziness are possible and have happened in the cultural sphere in recent years. As one case, the Mexican-born rock guitarist Santana won a record 9 awards at the Grammys of 2000 (the award for American popular music). Forty years ago this could not have happened, and it signals a major shift in attitudes in USA towards Mexicans. Santana's personal history shows how complex the movements of people across borders can be in an age which accepts such movements. Santana was born in Tijuana, in Mexico but close to the border, so he was already a Mexican who was not fully Mexican. His body crossed the border, and so did his mind and soul. He brought his skill with a guitar – an instrument US popular music borrowed without acknowledgement from Mexico – and became immersed in rock and roll, US music heavily influenced by African rhythms. His music as it has evolved and was rewarded with all these 'grammys' represents a fuzzy cultural mix, including musicians from both countries, black and white and brown, music from different cultures, lyrics that talk about unity and inclusiveness, songs in both languages. Such a fuzzy construction of multiple identities would not be comprehensible for any system of fixed identities (SFI), and it transcends the powerful but simplistic meanings of 'Chicano'.

The case of Santana illustrates two important paradoxes of human identity which describe the potential identities that can exist under globalisation (Dimitrov, 2001):

(1) an identity does not change in time and space but identities are constantly in a state of change.
(2) the stronger the connection between human identities, the higher the degree of their uniqueness.

The first paradox pivots around the Latin root of 'identity' from *idem* - 'the same'. An identity that remains the same inevitably becomes dogmatic and does not evolve or transform, just as an identity that constantly changes is transitory, with no ground and therefore vulnerable to external manipulations. *Na ca ço, na ca año* says an enigmatic statement written in Sanskrit more than 5000 years ago: *neither the same, nor different*. For instance Santana over his life has changed in many ways, but there is a bed-rock which is the same or continuous.

The second also uses the Latin root in a double sense. 'Identity' refers to ways in which a person is *the same as* social others – a Mexican like other Mexicans. Yet Santana is not like other Mexicans in his sense of strong connections with others in his complex identity – American, rock musician, worker for peace, spiritual person, etc. These other facets do not make him a weaker, less definite person, with less identity, but the opposite: more unique, yet more connected. Through the creative power of existence our inner selves remain inseparably connected with each other and the universe.

In this respect Santana is one of the products of a globalisation, which is also overseeing a massive erosion of local unitary identities. His specific identity is a 'fractal' version of the multiple possibilities for identity that now exist. This identity is not imposed on him by external forces but a creative achievement on his part. He won 9 Grammys for his songs. Perhaps he should have won a tenth, for his creation of 'Santana'.

6 Fuzziness of the West and the East

One of the secrets of the success of fuzzy thinking has been its attitude to natural language and the human thinking encoded in it. Where classic forms of science distrust human languages and the ideas that have evolved over many millennia, fuzziology is open to these rich traditions. Fuzziology recognises their value in their own human sphere. It can 'compute' with forms of wisdom as well as with perceptions and words. In this chapter our aim is not to rewrite the history of thought as a progression towards the climactic moment of fuzziology, nor to argue that fuzziology is just a new version of some very old ideas. The connections over time we see are full of jumps and discontinuities.

Fuzziology is no less original for having origins and antecedents, and will be no less powerful if it acknowledges them, and learns what it can from the past at the same time as it is obliged to go beyond.

In a popular book on fuzzy thinking, Kosko contrasted East and West, Aristotle and the Buddha, as representatives of black-and-white-thinking and fuzzy thinking respectively: 'in the West, Aristotle gave us our binary logic and much of our scientific world view' (Kosko, 1995;p. 68). This claim finds a secure place in a myth of Western thought which sees the ancient Greeks as the inventors of western rationality, the corner stone of modern science. So the author looks to the East, to the wisdom of the Buddha, of Daoists and others, as though it could never be found in the West until the time of Fuzzy Logic.

We believe this picture profoundly mistakes the course of world history of ideas by emphasising the stark contrast between 'East' and 'West' - a contract that is very widespread in the popular mind. Some may ask: does it matter? To which we reply that bad history can never be a good foundation for any kind of understanding of where we are in the present. In this case, fuzziology exists in the modern global world, where the words 'East' and 'West' still have powerful ideological effects. If fuzzy thinking is 'natural' to the East, how could people in the West dare to develop it, then sell it back to its original owners? And meanwhile back in the West, how can European fuzziology hope to compete against its own 'natural' (crisp) logic, built into 2000 years of history?

6.1 Fuzzy Dionysos

The 19th century philosopher Friedrich Nietzsche had a different account of the origins of the 'miracle' of the Greek culture. He captured the two main strands through two of the Greek Gods, Apollo, God of light, God of Reason, and Dionysos, God of orgiastic rituals. Apollonian man, he writes, is like the philosopher Schopenhauer's description of 'a man caught in the veil of Maya': "Even as on an immense, raging sea, assailed by huge wave crests, a man sits in a little rowboat trusting his frail craft, so, amidst the furious torments of this world, the individual sits tranquilly, supported by the principium individuationis and relying on it." (Nietzsche, 1956). In the Dionysiac orgy, he says, everything is different.

"Now the slave emerges as a freeman; all the rigid, hostile walls which either necessity or despotism has erected between men are shattered. Now that the gospel of universal harmony is sounded, each individual becomes not only reconciled to his fellow but actually at one with him - as though the veil of Maya had been torn apart and there remained only shreds floating before the vision of mystical Oneness" (ibid.; pp. 21, 22).

According to Nietzsche, the ancient Greeks knew both halves of this double truth; Apollonian reason, cutting things up, isolating and dividing things in order to understand and have the illusion of control, and Dionysiac fuzzy transcendence. The rites of Dionysos were Greek, but they came from Asia Minor Nietzsche says, and as far back as Babylon. And as he could have added, South as well as from the East, from Egypt, whose Osiris was the same as Dionysos, or from Crete, island of the mysterious Minoan civilisation which flourished a thousand years before classical Greece.

Dionysos and Apollo were the two Gods who presided over Delphi, 'the navel of the world' for the Greeks, their most sacred place. The oracles of Delphi were famous in the ancient world, prophesies that were always right because they were so ambiguous they could never be wrong. Apollo, God of light, was the source of the rationality of the oracle, but Dionysos was the source of its fuzziness. Dionysos was the ancient God of Fuzziness. The principle of fuzziness he represents was valued and worshipped by the ancient Greeks. It is not foreign to Western thought, but the contrary. It is one of its roots, one of the fountains that has nourished it from its beginnings.

6.2 Philosophy of Flux

The science, art and culture of the Greeks came from the East as well as the West: in many cases from the 'middle' east (the 'fuzzy' east). It is a myth that Greek rationality was a virgin birth, a miracle that suddenly appeared in Athens, at the centre of Greece, in the 5th century BC, without precedents or connections with the rest of the world. Bernal has criticised this myth as an example of racism Bernal, 1987). He sees it as the product of a conspiracy of European scholars, especially in the 19th century, trying to construct European rationalism as a purely European invention: made by the Indo-European Greeks alone without any help from the Asiatic others, Egyptians, Babylonians etc.

In fact the Greek miracle was constructed in a more complex way. Homer, greatest poet of the Greeks, was probably a product of Ionian Greece in around the 8th century BC, in Asia Minor not on mainland Greece. Greek philosophy was born in the Ionian colonies, on the dangerous edge of the Greek-speaking world, in what has been called the 'Ionian enlightment'. Thales of Miletus united science (he successfully predicted an eclipse in 585 BC) and philosophy, in which he proposed theory of the nature of the cosmos. Anaximandros, also of Miletus, refined this theory, proposing a scheme in which 'the Unbounded' contained opposites, such as hot/cold, which separated out to form kosmoi ('ordered entities', or 'systems'). Only slightly further afield, Pythagoras, born around 580 BC on Samos, an island which was part of the Ionian league, travelled for many years through Egypt and Assyria before finally settling down in Croton, Southern Italy, on the western edge of the 'Greek world'. From there he taught his doctrines, a mathematical theory of the universe (*all is number*) that has been influential on the development of science. Who knows, he probably did not invent 'Pythagoras theorem' but learnt it from the Babylonians or the Egyptians, as also π, but both were important in his synthesis.

6.2.1 Zeno

The philosopher Zeno, born around 495 BC, also taught in Italy, at Elea, a Greek colony set up by Ionian Greeks fleeing from the troubles in their homeland. He was responsible for four famous 'paradoxes', problems which Aristotle recognised were fundamental challenges to the principles of logic he was trying to develop. The 'sorites' paradox for instance starts from the idea of a heap of grain. If one grain is removed, it is still a 'heap'. If another is removed, it is still a 'heap', and so on, till finally the last grain is removed, each change the same size as the previous one which was not designated as sufficient to create a 'non-heap', with no moment when it is not a heap. So there are no grains left, but there is still a 'heap', or so the paradox says. The paradox turns around the inability of 'crisp' logic to cope with the concept of 'degrees of heapness'. Zeno in effect was projecting the need for fuzzy reasoning, by pointing out the absurdities that arose with non-fuzzy thinking. His example interestingly uses the same image as Per Bak used to illustrate the concept of 'self-organising criticality' (Bak, 1996). Bak imagined adding not subtracting grains to a pile or 'sorites' of sand, producing a particular form which was subject to 'avalanches of criticality', small or large slips that altered the shape of the heap. As in Zeno's case, the size of the difference that makes a difference is a fuzzy quantity. Zeno's paradoxes were taken very seriously by the ancient world, and deserved to be.

Ionia not Athens was where Greek thought was born, and then exported around the far-flung Greek world. As a site for intellectual production it had two notable qualities. One was its political instability. These Greek colonies spoke Greek but were small players in the power-games of this contested part of the world. After the Hittite empire collapsed between 1200 and 1000 BC, many contenders emerged. In the 6th century, for instance, Croesus king of Lydia, reputed to be the richest man in the world, was the main protector of these Greek states, and built the temple of Artemis at Ephesus, one of the 'Seven Wonders of the World' in ancient times. He challenged the emerging power of the Persians and their king, Cyrus, who defeated him in 546 BC, and tightened his hold over the Greek colonies. The triumphs of Ionian thought were produced in a constant state of war. At the same time, these Greeks were citizens of two cultures, with access to the rich intellectual traditions of the 'east', of Phoenicians, Syrians, Assyrians, Medes and Persians. In order to survive they had to be open to complexity, multiplicity, contradiction, paradox. In its most creative beginnings, 'Greek' thought was only Greek to a degree. It was both 'Western' and 'Eastern', a fuzzy matrix for the development of a philosophy that had many elements of fuzziness.

6.2.2 Heracleitus

All these tendencies can be seen in the work of Heracleitus, a major thinker in the Ionian tradition, whose ideas in many ways foreshadowed Fuzziology. He was born in Ephesus, home of the Temple of Artemis, the centre of worship of the Great Goddess in Asia Minor. The relevance of this to Heracleitus can be seen in his name, which means 'fame of Hera', Hera the wife of Zeus, 'Mother of the Gods' of Olympus. But Croesus, builder of the temple, was defeated 6 years before Heracleitus was born (546 BC). In 499 BC the Ionian states revolted against the Persians under Darius, but were defeated in 495 BC at Miletus. Many Greeks migrated west to Italy, but Heracleitus stayed till his death around 480.

Heracleitus mainly wrote about the cosmos, not about society, but he reflected on the problems of his society, urging the need for people to live together in harmony. In this respect, his ideas are built on the basic principles of social fuzziology, and this book needs to honour his name. All his key ideas have direct applications to social phenomena, and it is likely that this is no accident: this is where his impetus came from.

Like Anaximander he proposed a universal principle, which he called *logos*, the untranslateable word at the heart of fuzziology. His *logos* was a principle which saw *all things as interrelated*, in spite of appearances: hot and cold, health and disease, good and evil. If he had applied the principle explicitly to peoples, he could have added: Greeks and Persians, conquerors and conquered. Pursuing this principle of the resolution of opposites, he noted that one thing, for instance sea water, could be beneficial to fish but dangerous to humans. In systems such as human societies or ecologies, he argued that change in one direction will ultimately be balanced by change in the other direction. Even when things 'tend apart' they are actually being brought together, he said, proposing what in later ages would be called a cybernetic principle underlying social life.

Thales had proposed the single underlying basis of the universe was water, probably influenced by the Babylonian (and Egyptian, and Biblical) world view in which the earth is surrounded by waters, above and below. Heracleitus instead proposed that the single constituent of the universe was fire, 'everliving fire kindling in measures and being extinguished in measures'. This fire was an abstract principle, similar to the later proposal that energy in its permutations (between different kinds of energy, and between energy and mass). Like Thales,

Heracleitus drew his principle from religious cosmologies around him, in his case probably from Zoroaster's doctrine of Ahura Mazda, the divine light, also worshipped in a cult of fire, but the form he gave the theory was abstract and scientific.

Heracleitus believed that everything existed in a state of flux. One of his most famous aphorisms was *panta rhei*, everything flows, everything is in a state of flux, by which he meant that in spite of apparent fixity or stability, the reality is movement, faster or slower. 'Upon those stepping into the same rivers, different and ever different waters flow'. A river is an open system. It seems the same, with same name, but different are its waters, just as Ephesus kept its name whether ruled by Lydians or Persians or Greeks, always the same yet not the same. It is a very modern perspective on identity. It is also a kind of fuzzy thinking.

6.2.3 The Heretics

Ideas from earlier thinkers mingled in that part of the world over many millennia, back to anonymous thinkers from ancient Sumeria and perhaps earlier, and also forward. Eight hundred years after Heracleitus, Mani preached a religion which the Christians called a heresy, but was in fact an alternative religion, Manichaeism, which tried to bring together the ideas of Christianity, Zoroastrism and Buddhism into a single universal religion. Mani was killed by the Persian king, but his ideas flourished for two centuries, reaching Spain to the west, Beijing to the east, and Africa to the south. Persecution by Christians seemed to exterminate it from Europe by the 7th century AD, but offshoots of it kept springing to the surface, in the Bogomilists of Bulgaria in the 10th century and most famously in the Albigensians or 'Kathars' of Albi, in the south of France, who were brutally suppressed by Pope, who organised a 'crusade' to crush them in 1221. Those not killed in the war were systematically hunted out by the Inquisition, invented for the purpose of discovering 'Cathars'.

In instance after instance, these ideas were treated as dangerous to the state because they applied fuzzy interpretation to the concepts of religion, recognising contradictions in core beliefs, producing ideas of tolerance which were not what any ruler of any state wanted.

Equally important, in every case the repression of these ideas never destroyed them. The tradition was always skilful and flexible enough to go underground and survive, and emerge later when circumstances seemed more favorable. The Cathars disappeared as Cathars, but they remained a kind of secret society, dominating the literature of the early Renaissance, from the troubadors of France to the towering achievements of Dante, and also a major force in the development of science, brokers of the knowledge preserved in the Arab world. As with the Greeks, so with the Renaissance, Eurocentric scholarship minimises the essential role of 'the East' in the achievements of the 'West'.

6.3 Other Masters of Fuzziology from the West

In the first chapter we discussed the role of Socrates - the greatest master of fuzziology of all times. In this chapter we intend to reveal the significance of two other masters - Aristotle and Hegel

6.3.1 Aristotle

Fuzzy reasoning is often cast as an attack on Aristotle and 'Aristotelian logic'. In a limited sense this is true, if it refers only to the syllogistic reasoning associated with the name of Aristotle, based on the Law of the Excluded Middle - that A and not-A are incompatible. However, it is an extreme example of non-fuzzy thinking that makes a travesty of the work of Aristotle, who was one of the greatest intellectual figures produced by the Greeks, and who in many respects had nothing in common with so called "black-and-white logic" typical of the thinking of many of today's scientists and experts. Even the word 'logic' was not used in the works of Aristotle. The word he used was *analytics* - this is his own name for logic. The works on analytics are as significant for humanity as full of genius as his works on the philosophy of nature, on psychology, on metaphysics, on ethics and politics, on art and rhetoric.

In "The Categories", Aristotle distinguished expressions that exhibit propositional unity from expressions that do not; that is, he distinguished between a simple term and a composite statement that relates a subject to a predicate. He considered simple expressions *neither true nor false* and held that they may signify things in one or another of the following categories: substance, quantity, quality, relation, place, time, position, state, action, and affection. In chapter 5 of the Categories, Aristotle distinguishes within the category of substance between "primary substance" and "secondary substance." Primary substances are particular men, particular horses, particular stones, etc., and secondary substances are the species and genera to which the individuals belong. There Aristotle treated genus and species as substances of a derived kind.

How similar is this with what we study in fuzziology: *in order to grasp the fuzziness of an expression we must be able to transcend its appearance as true and false at the same time*, that is, to accept Aristotelean "neither true nor false". The primary and secondary categories reflect the fractal nature of complexity - the 'whole' of a particular man reflects the 'whole' of each human being belonging to the 'whole' of human kind.

Aristotle's treatise "On Interpretation" begins with a brief but influential discussion of the simple parts of sentences, such as 'names' and 'verbs'; it then considers complete sentences of various kinds and examines the analytical (that is, logical) relationships - contrariety, contradiction, implication - holding among them. The work also contains a pioneering account of 'modal' sentences ('It is possible that ...'; 'It is necessary that ...') and a celebrated discussion of 'future contingents': If it is already true that there will be a sea battle tomorrow, then how can the battle be considered a contingent event? For if the truth is already determined, surely the battle is fixed and necessary? Aristotle's answer to this is that certain types of sentences about the future are *neither true nor false*: again a direct connection with fuzziology's endeavour to avoid labels as 'true' and 'false' when considering any virtual unfoldment of reality.

The treatise "Sophistical Refutations" exposes forms of reasoning that appear valid on the surface but are in fact fallacious. Examples of fallacious arguments are 'begging the question', or circular argument (e.g., a 'proof' that the soul continues to exist after death because it is immortal); the 'fallacy of the consequent', or arguing from a consequent to its condition (e.g., 'if a man is a drunkard he becomes destitute; Peter is destitute: therefore Peter is a drunkard'); and the 'fallacy of the irrelevant conclusion', wherein, instead of proving the fact in dispute, the arguer seeks to gain his point by diverting attention to some extraneous fact. The delusive fuzziness is often based on expressions that 'appear valid on the surface but are in fact fallacious'. So Aristotle can help us (when applying fuzziology) to recognise specific kinds of delusive utterance.

Fuzziology with Aristotle removed would be impoverished. Aristotle without his insights into fuzziness would be diminished. In both cases, the price of a tidy intellectual scheme is far too high.

6.3.2 Hegel

In modern times, the first great challenge to what was called Aristotelian logic is associated with the theory of dialectic as elaborated by the early 19th century German philosopher Hegel. In thinking about the paradoxes of transcendence within social fuzziology it is impossible to avoid the giant figure of Hegel, and his legacy is still important, still controversial today after 200 years.

Hegel developed his theory of the dialectic process first in relation to the fundamental categories of thought, and then to the various disciplines of knowledge of his day. His starting point in metaphysics determined the form his thought took. This can make it seem strange for scientists and others today whose starting point is very different.

If someone tries to think the notion of pure Being (the most abstract of categories) they will find it is Emptiness, Nothing. Yet Nothing is. The notions of pure Being and Nothing are opposites, yet each passes into the other as someone tries to think it. But Hegel proposed a way out of this see-sawing contradiction: to reject both notions separately and to affirm them both together: to make being and nothing simultaneously to not be and to be. For Hegel this was the mediating category of Becoming, since what becomes both is and is not at the same time.

This process moves by a regular oscillation from positive to negative moments of thought to a moment that is both negative - a double negative - yet also positive. Later Hegelians called the terms thesis, antithesis and synthesis, but 'synthesis' is not an adequate way to describe the creative double negation, a moment of transcendence which Hegel called aufhebung a 'taking out' (negation) or a 'taking up' (transcendence). This itself becomes a new thesis which will be negated in a dialectic process that continues endlessly. Every later term in the dialectic is formed out of contradictions, and so is the first one, since it is unthinkable without reference to its opposite. Without contradiction, tension, conflict there would be no movement.

Hegel's tradition split into two main groups, 'old Hegelians', who maintained his conservative stances, and 'young Hegelians' who included Marx, who saw the doctrine of the dialectic as a description of the revolution which excited Hegel.

Hegel's achievement remains, as the starting point for the development of a new kind of logic, one in which A and not-A coexist in tension, producing not a fuzzy compromise but an unpredictable new form - neither fuzzy nor non-fuzzy: neither (A *and* not-A) nor (A *or* not-A). The full development of this new logic is a major task for fuzziology, and social fuzziology should be its laboratory.

6.4 Myths of Fuzziness

6.4.1 Wisdom of the Ancient Myths

There is a danger in following the ideas of great thinkers that the embeddedness of ideas in the everyday language and thought of ordinary people will be lost. Yet it is one of the essential features of fuzzy thinking that it does not dismiss or ignore the language and thought of ordinary people, but on the contrary uses it as a resource. In the case of the Greeks, alongside the works of major thinkers whom few read, even if they were as famous as Aristotle, there is a stream of familiar images which carried the principles of fuzziology then, and continue to do so now. We will illustrate with a number of popular phrases, associated with myths, which are still known and meaningful today, in contemporary society in all modern European nations.

6.4.1.1 Procrustean Bed

Procrustes ('the Stretcher') was the name of a robber in a myth about the hero Theseus, journeying towards Athens where he would ultimately become one of the wisest of kings. Procrustes waylaid weary travellers with the offer of a bed, but his bed was not an ordinary one.

Instead of 'one size fits all' he had two sizes that fit no-one: a short one for tall men, whom he cut down to fit it, and a long one which was a rack which stretched short men out till they fit it. Wise Theseus killed him and broke up his beds.

It is as clear a metaphor today as it was in ancient Greece: a lesson on behalf of fuzziology for would-be kings or modern managers, or anyone whose will to power takes the form of fixed categories and control at any cost.

6.4.1.2 Gordian Knot

The Gordian knot is part of a story told about another great King, Alexander the Great. The original knot was tied by Gordium, a legendary king of Gordia, a city in Phrygia in Anatolia, strategically a key route between the Middle East and Asia. Gordium was a peasant who came to the throne by miraculous means. He was driving his ox-cart to the city when a royal eagle, bird of Zeus, perched on its pole and refused to fly off. Gordium was delighted with the omen, and drove on to the city to make an offering. He met a young prophetess at the city gates, who agreed to help him make the offering. He proposed marriage to her and she accepted, and the pair drive into the city together on his ox cart. Meanwhile in the city the old king had just died, and an oracle announced that their new king and his queen would enter the city driving an ox cart. Gordium and his bride, accompanied by the eagle, fitted the prophecy, so he was made king, and she his queen. In gratitude he dedicated the ox cart to Zeus and left it in the Acropolis of Gordia, tying the yoke and the pole together with a special knot, the 'Gordian knot'. The oracle announced that whoever discovered how to undo the knot would become master of all of Asia.

There are a number of stories about how Alexander solved the problem. The most popular is that he simply cut it through with his sword. This would make the solution a triumph of brutal linear logic of power overcoming complexity. Another story is that he pulled the pole out, so that the knot collapsed and could be easily unravelled. Either way, Alexander the Great, Aristotle's most famous pupil, adopted a solution that showed 'lateral thinking' to use De Bono's well-known term, and he did become master of Asia, as had been prophesied.

There may be a deeper level to this myth. In Gordium's myth the pole (*axion* in Greek) and the yoke were metaphors for the two poles at right angles that organised the cosmos, so that the knot symbolised the complexity that held the cosmos together. His challenge to his successors was to find the point of balance, the fulcrum, the cosmic axis. Only Alexander succeeded in doing this, just as his conquest of Gordia, militarily and symbolically, did in fact open up Asia to his conquering armies. On this interpretation, the Gordian knot stands for breaking with convention, finding the pivotal point and thus solving a problem that seems insolvable. In this sense, the myth of the Gordian knot reflects the power of the transcendent logic of fuzziology in dealing with complex situations.

6.4.1.3 Achilles Heel

The myth of Achilles' heel is the counterpart of the Gordian knot. This myth is part of the English language: the 'achilles tendon' is the tendon behind the heel. Achilles tendons are one of the most vulnerable parts of the leg for a champion runner, whose superbly firing muscles and months or years of training will be rendered vain if this small piece of tissue tears. The 'Achilles heel' is the point of vulnerability for a system, where even a minor intervention can cause the system to crash.

Achilles was famous as a fast runner, and no doubt the ancient men who were models of the mythical Achilles knew of this danger all too well. The myths give another explanation. Achilles was the son of a goddess, Thetis, and a mortal, Peleus. He was therefore 'fuzzily' both mortal and immortal. According to one story, Thetis wanted to make him fully immortal, so she took him as a baby by day to Olympus, where she anointed his body progressively with ambrosia, the magical food of the gods, so that it would be immortal, while by night she put him in a fire to burn away the mortal parts. Peleus discovered what she was doing when only the ankle bone was left unprotected and charred. He took the baby away with him, and replaced the ankle bone with one from the swift-runner Damysos in the first recorded 'transplant operation' in history. Achilles grew up invulnerable to all wounds, except to his heel. He died in the Trojan war because Paris, his enemy, knew of the weakness, and an arrow guided by Apollo struck him on the heel and he died in agony.

In another version of the myth, Achilles was invulnerable in every part except for his heel because Thetis dipped him into the river Styx, the river of death that protected him from death. But she had to hold him somewhere, by the heel, so this one part was not invulnerable. The outcome was still the same: he died in battle to a wound on his heel.

In the first case the meaning of the myth is that the transformation of a mortal (imperfect, mixed) system into an immortal (perfect, unmixed) system cannot be achieved without a fuzzy residue of elements that resist or escape purification.

These elements, however small or insignificant they seem, will still have the capacity to negate the rest of the system. In the second case, the reason the imperfection is there is even more paradoxical. It is the hand of the Goddess herself, acting in a human way on the human body of her son, that creates the site of vulnerability. The contradiction is not only in the baby, part human and part god, but also in the divine component, in this case the goddess, who acts in ways like a woman. Her hand and his heel contain the fuzziness of the two intersecting systems in their acute form, 'fractal' versions of the fuzzy union of human and divine in the man and his goddess mother.

The two versions do not contradict each other. They coexist as two different reflections on the vulnerability of almost perfect 'anti-fuzzy' systems, and the risk on the way to perfection. A modern version of this myth is the episode in Star Wars where Luke Skywalker destroys the almost invulnerable Death Star by dropping a bomb down its only point of weakness: a passage so small that the designers hardly took account of it, the Empire's 'Achilles heel'.

6.4.1.4 Pandora's Box

Not all ancient knowledge is equally wise. To illustrate the point we will take the well-known myth of Pandora and her box, another phrase which is part of the English language. In the myth, originally written by the poet Hesiod in the 8th century BC, Pandora was constructed by Zeus as the most beautiful but foolish woman who ever existed, and given to Epimetheus ('After-thought') as a trap. At first Epimetheus politely declined the gift, but later, fearing Zeus would be angry if his gift was rejected, he married her and fell in love with her beauty. In this myth Zeus is not a benign god but the contrary. He is hostile to humanity, angry with Prometheus ('Forethought') for giving humans the prerequisites for civilisation, including fire. Prometheus, in his attempts to improve the lot of mankind, had also captured the ills that plague humanity, such as sickness, old age and insanity, and put them all into a box or jar, which he warned Epimetheus never to open. But Pandora, of course, an idle and curious woman, opens the box, and out fly all the plagues that were contained there. So Pandora's Box (which was really Prometheus's box) is a symbol of something that should not be opened because once that has happened, all hell will be let loose.

But Hesiod was a misogynist, and so is the Zeus he invents. In Hesiod, Zeus constructs the stereotype of woman that women have protested against for many centuries - very beautiful and dangerously foolish. Hesiod's Zeus not only gives 'After-thought' this dangerous gift, compelling him to marry her not relying on her natural charms, but he suspects 'Forethought' of being the lover of Athena, goddess of wisdom on Olympus. Athena does indeed help 'Forethought', but only because like him she believes humans should be helped to become more civilised, and those who work towards this end should be praised and supported, not punished. Hesiod has also twisted the nature of Pandora, whose name means 'all giver'. This is the name of the Great Goddess, the bountiful giver of all things. The real Pandora did not let out all and only bad things. On the contrary, her 'box' was her all-creating womb, the source of everything that is.

Paradoxically the myth works better when the box is restored to Prometheus, 'Forethought', the well-meaning god of plans and schemes. He gives humanity fire and other gifts of civilisation, but he also attempts to remove and contain the bad things which are equally the products of 'Pandora', so fuzzily intermeshed with the good things that it is impossible to separate them. So Prometheus represents the dangers of non-fuzzy thinking even when, or especially when, it is accompanied by the will to good. Hesiod's story gives the impression that the ills in the box, which were taken out of everyday circulation, now re-enter the world with a greatly increased capacity for harm.

This lesson from Pandora's box applies also to all uses of the ancient wisdom in a fuzzy framework. If all the 'good bits' are taken to one side and stitched together into a single image, as Hesiod's Pandora was constructed by the gods out of the beautiful parts of womanhood, then the result will feel empty and foolish. The collection of the 'bad' bits, separated, confined and constrained, will not be eliminated, and they will become more dangerous in that form. The 'good bits' will still need the 'bad bits' to form the springing board for transcending the fuzziness of duality embedded in human thinking, and thus to reconstitute the original infinity of the real Pandora and the eternal Promethean urge for creativity.

6.5 The Fuzzy Wisdom of Tao

In deconstructing the opposition that is made between the 'East' and the 'West', it is important to point out the eternal fuzziness of the West. But that should not be at the cost of denying the long, rich tradition of fuzzy thought in the East. We already discussed the greatness of the Vedic thought developed thousands of years ago.

Each tradition in the treasury of the Eastern thought is itself so ancient, with so many forms, that any attempt to describe them all would be in vain. Instead, here we will illustrate the point by reference to one famous text, the Chinese *Tao Te Ching*.

Chinese philosophy today flows from foundations that were laid down in the 6th century BC, characterised by a dialectic between Confucianism and Taoism. Confucius (Kongzi to the Chinese) was born around 551 BC, about 6 years before Heracleitus, and died in 479, one year after Heracleitus. The supposed author of the Tao Te Ching, Lao Tzu, is said to have been born at much the same time as Confucius, though the book itself and its Taoist doctrines are usually seen as a collection made over centuries by many nameless hands. The similarity of dates and doctrines of the Chinese works to those of the 'Ionian enlightenment' of the West is a salutary reminder, that the non-fuzzy distinction between East and West is bad history.

Confucianism and Taoism were produced in difficult times, as was the philosophy of the Greeks. Confucius and Lao Tzu struggled to survive in the court of the Zhou kings of north eastern China, a dynasty that had come to power in the 12th century BC and was to be defeated in 221, after a long, turbulent period known as the 'Warring States' by the Qin dynasty, whose king became the first emperor of China. In their different ways, Confucians and Taoists aimed to provide advice to the kings and rulers of their day, though they also tried to preserve some distance. This is obvious in the case of Confucius, but it is still apparent in the more mystical-seeming Taoists. Both were social theorists as well as philosophers.

Alongside the Confucians and the Taoists, there was another group, who need to be mentioned as an antidote to simplistic ideas of the total difference between the materialist West and the idealistic East. The 'Legalists' came somewhat later than Confucius, but their intent was to provide the sort of advice rulers really wanted and needed: 'What the world now calls righteousness is the establishment of what people like and the abolishment of what they dislike...Punishment produces force, force produces strength, strength produces awe, awe produces virtue. Virtue has its origin in punishment'. This is similar to Machiavelli in the West, in the 16th century, though the argument that might is right and 'virtue' is a convention was known in ancient Greece in Plato's day, before 380 BC.

The Tao Te Ching (Mitchell, 1996) begins with a famous paradox, which pivot around the fuzzy immensity of the key word 'Tao' (or 'Dao'):

There are ways but the Way is uncharted;
There are names but not nature in words.

All translations into English (or other languages) have difficulty translating these words because they exploit and create all the contradictions inherent in this word, in similar ways to the use of logos, the 'Word', in St John's Gospel. A modern dictionary lists many alternatives: a way, path, road; a method, doctrine or theory; words, thoughts or beliefs. This ambiguity in the meaning of the word, which is typical of all human languages, is then filtered through a syntactic form that magnifies the potential ambiguity. The syntax of classical Chinese is a machine for refuzzifying language: not (as is the case with modern English, and modern spoken Chinese) a source of fuzzy constraints. The actual words are Tao ke tao, fei chang tao; ming ke ming, fei chang ming: literally the 'tao' that can be 'taoed' is not the eternal 'Tao'; the name that can be named is not the eternal name.

In classical Chinese, the complex symbol, in this case 'Tao', can be singular or plural, noun or verb without change of form. Ke, can, signals that the second Tao is a verb, a potential action in this case (fuzzily both an action and not an action) but it is not clear how the thing *tao* (way, road, path; method, principle) can turn into an action, or who is doing it. But as soon as it is turned into this highly fuzzy action it is instantly negated, becoming the transcendent *neither* (A or not-A) *nor* (A and not-A). The 6 words of the line follow a triple movement: assertion (tao): fuzzy statement (ke tao): contradiction (fei chang tao) which then loops back to the first assertion, turning it into an impossible statement, a term whose meaning grows ever richer as it is unable to achieve closure around any determinate meaning.

The book then goes on to insist on the existence of what is negated, what does not exist:

Thirty spokes are united around the hub of a wheel,
but the usefulness of the wheel

depends on the space where nothing exists.
Clay is molded into a vessel,
but the usefulness of the vessel
depends on the space where nothing exists.
Doors and windows are cut out of the walls of a house,
and the usefulness of the house
depends on the space where nothing exists.

Therefore take advantage of what exists,
and use what does not exist. (11)

This conception of a material nothing that has material effects and is therefore something is pointing in the direction of physics and mathematics, a new science. The same law of contradictions is also directed to social categories:

Be aware of your masculine nature,
But by keeping the feminine way,
You shall be to the world like a canyon,
Where the Virtue eternal abides,
And go back to become as a child.

Be aware of the white all around you,
But remembering the black that is there,
You shall be to the world like a tester,
Whom the Virtue eternal, unerring,
Redirects to the infinite past. (28)

The first part recalls another important part of the ancient Chinese wisdom, the yin-yang doctrine, which insists on the need for a harmonious balance between yin and yang in individuals and in the cosmos. In their most basic meaning they are female and male, but also refer to a cosmos understood in these binary but reciprocal terms: dark and light, moist and dry etc. These are similar pairs of opposites as Heracleitus saw united in the logos. In both traditions, fuzziology applies equally to people and the universe, so that their fuzziology is always social fuzziology.

However, the ancient Chinese like the ancient Greeks were actually very sexist in their social practices. Women in both societies were excluded from most professions, confined to the domestic sphere, with restricted rights and open to oppression and exploitation. Lao Tzu addresses his words to a man, of course. He urges him to be aware of his 'feminine side', and assumes no woman would be listening to him, needing to be aware of her masculine potential. As in the case of Hesiod, the Ancients were not disembodied beings outside the ideologies of their times, and their wisdom carries many traces of these one-sided views without ceasing to be wisdom.

The Taoists also had political advice for their rulers, which are still applicable today:

The world is won by refraining.
How do I know this is so?
By this:

As taboos increase, people grow poorer;
Where weapons abound, the state grows chaotic;
Where skills multipy, novelties flourish;
As statutes increase, more criminals start. (57)

The Taoists learned from the Legalists and vice versa to understand the actual political scene deeply. As well as being 'mystics' they were political scientists able to see politics as a chaotic system in which actions can produce exactly the opposite to what is intended. These paradoxes have a similar form to Zadeh's Principle of Incompatibility, to which these statements could be appended as corollaries or consequences. It is fair to say that Taoists laid down the basis for social fuzziology more than 2000 years ago.

6.6 Aboriginal Wisdom

Today's ecological problems are created by all of us and we are the beings who have to deal with them. It is an illusion that science and technology, no matter how powerful, can save the world from ecological disasters. The idea of scientific and technological 'miracles' creates a distorted image of human power that 'we are beings who possess nature and that our ability to transform and model it, using the power of our omnipotent brain, will help us to deal successfully with the ecological crisis'. Such an attitude is totally wrong because it considers human beings as being outside nature, imposing their decisions and actions upon it. By putting ourselves in a position of dominance over nature, we disenfranchise ourselves completely, because we are part of this nature. The way we relate to other parts of it is of crucial importance for our survival (Russell, Dimitrov and Fell, 1994).

Nature is considered as a force that must be kept at bay by human will and strength. Such a point of view justifies easily the intensive exploitation of the earth's natural resources and of all other creatures in the name of satisfaction of the human requirements. The same point is visible in all approaches, applying to the design of our contemporary military, government, economic and even ecological (!) systems.

The more we consider ourselves in a particular pre-selected position to nature, as the most powerful beings created to rule and manipulate it according to our needs, the less powerful we are to deal with our ecological problems. The transcendence of this dramatic paradox lies in admitting that the laws and principles from which the natural world arose are the same as those that generate human culture and society, and that we have equal opportunity to exist and evolve with all other creatures. The Aborigines of Australia have had this understanding for a hundred thousand years. It is no wonder that a similar understanding is typical for indigenous people who live nowadays in other countries - for example, Indians in America, Maoris in New Zealand, etc.

"It is the land which possesses us - we do not possess it" - this is the leitmotiv of Australian Aborigines' culture. We are created by nature to live in harmony with it - "we have been here since the time before time began; we have come directly out of the Dreamtime of the Creative Ancestors; we have lived and kept the earth as it was on the First Day" say the Aboriginal Tribal Elders. The same forces which permanently pulsate in Nature, making the volcanoes erupt, the tides flood, the planets revolve and the sun shine, pulsate in us. "The Aborigines refer to these forces as their Creative Ancestors. Everything was created from the same source - the dreamings and doings of the great Ancestors; all creatures - from stars to humans to insects - share in the consciousness of the primary creative force, and each, in its own way, mirrors a form of that consciousness. In this sense the Dreamtime stories of Australian Aborigines perpetuate a unified world view" (Lawlor, 1991). This is a 'unified world view' such as complexity science appeals to when dealing with humanity's problems. It also is essentially fuzzy in the basic categories that are needed to think about it.

The idea of unity, penetrating deeply the naturally born Aboriginal 'complexity thinking' compelled them to respect and adore the earth as if it were a book imprinted with the mystery of the original creation. The goal of life was to preserve the earth, as much as possible, in its initial purity.

How can we be pure, if our surrounding environment is polluted? "I feel it with my body, with my blood. Feeling all these trees, all this country... when the wind blows you can feel it. Same for country... you feel it. You can look, but feeling...that put you out there in open space", says a Kakadu Aborigine. Like the human body, the country is considered nonsegmentable; there are distinguishable features such as thighs, abdomen, and chest, but they form integral parts of a continuous living being.

The present ecological effort to re-establish a bond between human societies and the natural environment seems superficial compared to the Aborigines' deep identification with nature. For thousands and thousands of years this deep connection appears to have preserved both Aboriginal culture and the local environment in a harmonious balance. Without romanticising and idealising this culture, perhaps we can seek to understand their degree of identification with nature as a guide to transforming our present relationship with our environment. Otherwise we will live with ecological crises that threaten our very survival.

In addressing these issues, a practical step might be the use of social fuzziology to establish a common language concerning environmental issues between different sectors of society - farmers, technologists, academics, students, etc. The second step would be to practise speaking the language of nature: a natural language, which *goes beyond* our current languages of science and the environment. Once upon a time we must have spoken such a language better than we do now, but it seems we have forgotten it. There are aboriginal people who still keep its secret and who could help us to reclaim it: if only we know how to ask.

Traditional Aboriginal knowledges are still carefully preserved by authorised guardians, Aboriginal elders who have gained the right to know and pass on it on to selected others. These people are experts, and to build in the expertise of experts, and the language they use to encode it, has been one of the successful strategies in devising fuzzy expert systems. Although it may go against the grain for some highly educated citizens of the modern world to recognise it, this Aboriginal expertise needs to be built into an expert learning system for dealing with the ecological crisis.

6.7 Using Fuzziness of the Ancient Wisdom

The ancient wisdom could serve as an effective medium for metaphorically conveying complex concepts and principles to a broad spectrum of people. In this section we show how a fuzziological approach can be used to translate the metaphoric language of ancient myths and legends into a 'soft' scientific paradigm which helps us to understand better the social complexity of our time.

The wisdom of humanity combines in its indestructible integrity both the rationality and irrationality of human thinking and acting. Human reason is the torch of rationality in the darkness of unknown and uncertainty. Myths, legends, metaphors and paradoxes are the multifaceted eyes of irrationality - through them we can see in the darkness of unknown and uncertainty without any other torch. On the boundary between the rational and irrational is *fuzziness*.

Human beings are 'at root' irrational (Williams, 1988) - the hidden power of their illogical intuition challenges constantly the constraints and 'norms' of rationality. But far from being an enemy of reason, it can be a powerful ally. Any discovery, before it happens, always seems irrational - it challenges and contradicts rationality. After it happens, rationality adopts it immediately - it becomes part of a new rational explanation. Just at the border, just at the edge of this transition, is the zone of fuzziness.

How often does that burning feeling of an approaching insight bring us disquiet: the old is not accepted any more, the new is felt almost in certainty - but when we try to articulate it in words it seems so unclear, so imprecise, so fuzzy, that rationality does not tolerate it, but attacks it, rejects it.

Fuzziology tolerates the 'conceptual twilight'. We can say 'Yes and No' at the same moment, and move beyond their duality, we are at ease with 'More or Less', 'Maybe' and 'Possibly' in our explanations. Linguistic paradoxes do not bring us to frustration or passivity, but on the contrary - they help us to see better what is beyond the parallel use of opposites, to understand the 'game of words', and make a leap into another space where what was heretofore known as 'truth' dissolves into shadows. The shadows interweave, interlace, intertwist, and from the fuzzy symplectic ball they build emerges the *new*.

Fuzziology helps us to penetrate into the wisdom of ancient myths, legends, fables and metaphors. The situation described in a myth is generalised, put into a broader context that relates to a spectrum of real life situations at the same time.

6.7.1 Bridging Myths and Reality

We can outline a kind of *bootstrapping algorithm* to move beyond the duality of the 'rational' and 'irrational' explanations of complexity, drawing on the fuzziness of the ancient myths or writings. The algorithm (Dimitrov and Dimitrov, 1994) includes the following steps:

- Generate rational explanations of different aspects of the studied social process; these explanations represent only partial descriptions of the process, as the complexity of it is incompatible with any black-and-white rational description.

- Select partial descriptions which do not contradict (harmonise, complement, clarify) one another.

- Combine the non-contradictory descriptions into a reasonable general description; it builds the rational part of the complexity model.

- Select partial descriptions that contradict one another.

- Generate various combinations between the contradictory descriptions, that is, combine them together in some common (quite possibly paradoxical) descriptions: they build the irrational part of the complexity model.

- Look for an ancient myth (legend, fable, metaphoric story) or statement (sentence, proverb, saying) compatible with some of the above irrational descriptions and roughly estimate the degree of their compatibility.

- If the degree of compatibility is 'high enough', the myth(s) or statement(s) is(are) adopted as operational holistic explanation(s) of the irrational part of the model.

- If no appropriate metaphor is found, the algorithm is recurrently executed starting with step 5, until appropriate metaphors are found or the time for search expires (in the latter case, the model is represented by its rational part only, found at step 3 of the algorithm).

- Both the rational and irrational parts of the model are used as a catalyst for emergence (creation, discovery) of an integrated (holistic) description of the studied social process.

Let us illustrate the above algorithm by an example related to the complex social process of human health.

6.7.2 Example

We are witnessing an ever-increasing use of medical drugs in the world (and hence a huge increase in the profit of the large pharmaceutical corporations).

The rational explanation is rooted in our belief that it is the medical drugs that cure our diseases. Another rational explanation is that the medical drugs strengthen the self-healing capacity of our organism. These explanations are supported by the majority of the medical practitioners and also by the medical drug producers; it is clear that strong economical interests are behind this support.

The irrational explanations, that is, the explanations, which are rejected by the contemporary medicine, are the opposite ones: the medical drugs cannot cure the diseases and they impede the realisation of the self-healing potential of the human organism. The medical drugs have only partial and temporary effects on the disease, as they deal only with the apparent symptoms of the disease and not with its source and with its dynamics, that is, with the forces and energies which are behind the holistic emergence of the disease. The more intensively we use medicines (and one cannot help but keep using them, as the effect of each dose is transient), the more addicted we become and the stronger the numbing effect that the 'curative' chemical substances exert on our self-healing potential. Eventually, the addiction results in a total lost of the self-healing capacity and full dependence on the medical drugs.

The ancient myth which suits the above irrational (paradoxical) situation is *the ancient Greek myth of Sisyphus*, a king of Corinth, condemned to roll a heavy stone up a steep hill, only to have it roll down against when he reaches the top. Like Sisyphus, we 'work hard' - pay a lot of money for medical help and drugs - to get rid of a certain disease by the use of a large amount of prescribed medicines. When 'at the top' we realise that another disease have appeared in the organism because of the intensive use of the drugs and our health 'goes down' again; then we 'work hard' to use a new, possibly larger amount of drugs to fight the new disease, only to realise after a while that we are sick again.

The *go-beyond* approach of fuzziology helps us to combine and then seek to transcend both the 'rational' explanations (that the medical drugs cures the diseases and strengthens our self-healing capacity) and their 'irrational' negation.

In this case, an emergent understanding may result in connecting the massive use of medicines in today's society with the serious ecological degradation of the natural environment. The most powerful holistic curing factors of nature: air and water, sunshine and forests, soil and rains, fruits and vegetables are not healthy any more. This affects negatively our self-healing capacity, and we rely on medicines to help us. As we discussed in the chapter 4.7, the self-healing capacity is a holistic property of the organism and only holistic factors can support it. The medical drugs are not such factors; it is the healthy nature that provides them. Therefore, together with the endless rational and irrational debates about the use and abuse of the medical drugs, urgent practical actions are needed to stop killing the Mother Nature (and ourselves together with her).

7 Key Terms in the Language of Fuzziology

Fuzziology like every other field of thought has its key terms, and this chapter will outline some of them.

Scientific language aims to have:

- maximum denotation (reference), minimum connotation (associations)
- crisp boundaries (no overlap with other terms)
- hierarchical (hypotactic) organisation (part of ordered set of terms)
- single principle
- no history.

The fuzzy forms of natural language make use of the opposite qualities:

- large set of active connotations
- fuzzy boundaries with related terms
- 'heterarchical' organisation with networks across different levels
- more than one principle, often contradictory
- active, relevant history inscribed into the field of meanings.

Both these potentialities of language co-exist, and have probably always co-existed. The important difference between the two is that the first has been prized too exclusively over the past 400 years, and the second not recognised as a productive form of language in its own right. In psychology Vygotsky described its evolution in the development of child thought, and called it 'complexive thinking' (Vygotsky, 1987). In philosophy Wittgenstein called a similar quality in language 'family resemblances', by which he meant the way that words and concepts in use tended to be organised through patterns of non-exclusive similarities and differences, in the same way that different members of the same family can look a bit like several or many other family members (Wittgenstein, 1994).

Modern linguistics has not developed an adequate theory of the fuzziness of language, and in fact is often opposed to it. This is largely because this form of linguistics follows a tradition known as structuralism, which is based on the ideals of the 17th century Royal Society's attitude to language. But before structuralism achieved its almost total dominance in linguistics there was another tradition of

19th century scholarship, which had more of a place for fuzziness: a form of study of language known as 'Comparative Philology'.

This approach has such valuable things to say to fuzziology that we will discuss it at greater length. This approach to the study of language began in 1898 with Sir William Jones' classic demonstration that Latin, Greek and Sanskrit were cognate languages, descended from a common origin he named 'Indo-european'. He hypothesised that Germanic languages, including English, and Celtic languages, were also descended from this common parent. Over the next 100 years his insights were vindicated many times over. The science of language built on its basis was one of the great intellectual achievements of the 19th century, influencing Darwin himself among many others.

For a variety of reasons too many and complex to be discussed here, comparative philology did not continue its dominance into the 20th century. One important factor was the work of Ferdinand de Saussure, the greatest exponent of comparative philology of his day, who nevertheless announced the death of this field in his posthumously published "Course in General Linguistics" of 1917 (Saussure, 1974) insisting on the need to replace it with structuralism, which became the dominant form of linguistics for the rest of the century. In USA the structuralism of Bloomfield of the 1940s was replaced by the structuralism of Chomsky in the 1960s, which has spawned rival schools, but all are equally structuralist.

It is important to recognise that this triumph of structuralism did not involve any demonstration that the triumphs of comparative philology were wrong. On the contrary, comparative philology explained everything that came within its sphere, demonstrating the affiliations of all Indo-European languages so convincingly that there was no important task left for good young scholars to undertake. Linguistics simply abandoned comparative philology for structuralism, because it seemed more scientific, not because it was in any demonstrated way more right. The monuments of 19th century scholarship were left without peers or successors.

In the second half of the 20th century, however, some of the principles of comparative philology returned in leading-edge humanities research. Raymond Williams, one of the founders of Cultural Studies, wrote a book, "Keywords", which became an essential text for the field, a guide to the irreducible fuzzy complexity of all the key terms in cultural studies, drawing extensively on their etymologies. Michel Foucault, pioneer of discourse theory, one of the most influential strands of analysis of language in cultural studies, called his method a genealogy. His genealogical approach included a recognition that history is constituted by ruptures and breaks as well as continuities, leaving traces in the form of contradictions.

In what follows we will illustrate these fuzzy conceptions of language in our descriptions of the irreducible fuzzy complexity of all these key terms. The starting point will usually be a reference to their history, drawing on their etymology, but in every case this is not an aim in itself, but a way of understanding the living presence of deep history in every aspect of language, not as something to be rooted out as far as possible but on the contrary, one of the precious resources the human past offers to the present and the future.

7.1 Fuzziness Defined

'Fuzzy' has a number of meanings listed in the Oxford English Dictionary (OED), but Zadeh's use of it, in 1965, has pride of place as a new meaning. Although Zadeh took the word from the resources of English, and his usage taps roots that go deeper than even he was aware, his version of the word is not the same as any of these. 'Fuzzy' in 'fuzzy sets' does not mean exactly the same as 'fuzzy' in any other context.

The success of this term is mysterious but not in doubt. There are many possible synonyms, in English and other languages, but none has the reverberations that have made 'Fuzzy' so potent a banner for this new form of logic, this new kind of thinking. For instance 'vague logic' is more or less equivalent, but much less interesting. In Spanish, it is translated 'lógica difusa', which misses the essential untranslateable quality of 'fuzzy'. 'Fuzzy' is untranslateable because it is itself fuzzy: a unique and unconscious concatenation of meanings from an unknown history.

Most etymological dictionaries derive this word from a Low German word *fussig*, meaning 'loose, fibrous, spongy' (Partridge, 1966). However, there is no close connection between this set of meanings and the word when it first appears in English, as cited by the OED, where it is applied to a kind of fungus ('puffes, fuzbals or fuzzes'), in Holland's 1601 translation from Pliny's Natural History. In

1616, Swift and Markham talked of 'soft mosse, or fuzzie grasse'. In 1685, West wrote that he 'came home well fuz'd'. In 1711 Swift applied the word to mental processes: 'I think I'm in a fuzz, and don't know what I say'. In 1720 Prior applied it to hair: 'that high fuzz of hair'.

This list shows the word had a wide range of meanings at its earliest appearance in English, and most of these are not close to the meanings of *fussig*. These two facts suggest that the root it comes from must be a common word, from a core vocabulary, with a long enough history to allow the generation of such diversity. This makes the link with *fussig* unlikely, as the sole and sufficient explanation of the word. It is necessary to look again for possible candidates.

The most obvious possibility is a Latin word which sounds very like it: *fusus*, the past participle of *fundo*, whose adverbial form is fuse, pronounced 'foozay' (White and Riddle, 1876). *Fusus* has as many meanings as *fundo*, which is a very important word, with many different senses, and a long history of its own, back to Indo-European roots, to **gheu(d)*, meaning to flow or pour, or run in a stream, especially in religious contexts (by convention, an * indicates a hypothetical root). Sanskrit has *hu* to worship with oblations, either liquid or burnt offerings. In Greek it became *cheo* (Liddell and Scott, 1840), originally to make an offering to the Gods, especially wine or other liquid, and other meaning signifying liquidity: to melt, to mingle liquids, to pour out, to cast metal. It also signified other things that were not liquid: to cause to fly out, as with sparks. From *cheo* come many meanings into modern English, including 'chemistry' (via the Arabic 'alchemy', the science of flows, mixtures and transformations, combining ideas of liquidity and heat.)

The Latin *fundo* and its participle *fusus* (meaning an object that has been acted on by whatever process is involved in *fundo*) has a similar range and complexity of meanings: to pour out, to melt, to cast metal, to relax the bowels; to pour out on the ground, hence extend: to speak at length, with excessive flow; or with non-liquid contexts, to throw out, cast wide; or to produce prolifically, as with offspring.

Latin was the second language for educated English people in the 17th and 18th century, when 'fuzzy' was born with its diversity already fully formed, as it had been for several centuries previously throughout Europe, when 'fuzzy' may have been used without its identity being marked strongly enough to catch the attention of lexicographers. The OED gives its core meaning as 'loose, volatile matter; mass of fine light, fluffy particles', which is one of the strands of meaning of *fusus*. But it also applied to liquidity, or to its effects; things which were full of water, which merged with each other like liquids, or like metals.

If this history is broadly correct then 'fuzzy' comes from a word (*fusus*) that refers to fire and water and their effects, to energised particles as well as liquids: to a world with unstable outlines, a world in flux. The family of English words that descend from it reflects the range of meanings of *fusus*. They include 'infuse', 'con-fuse' and 'transfuse' from 'melted or joined' with 'de-fuse' part of the same branch at one remove; diffuse' from 'spread out, extended' and 'profuse' and 'effuse' from 'pour out in abundance'. All these words are formed by the addition of a prefix to 'fusus', to limit or constrain the fuzzy range of meanings of *fusus* to a more specific (but still somewhat fuzzy) meaning.

It was relatively late in this history, in the late 17th or early 18th century, before 'fuzzy' was applied to the mind as well as the external world, to describe the fluid mind that saw the world that way. By this time its range of meanings was laid down, waiting a century to be taken up by Zadeh to describe a kind of logic that dealt with a fluid world.

7.2 The *Logos* of Logic

'Logic' also comes from a complex origin, but this time its history is not in doubt. It comes from the Greek *logos*: *logike* is an adjective which went so often with *techne*, art, that on its own it was understood as *techne logike* the art of whatever 'logos' means.

Logos itself is untranslateably rich, a very fuzzy word. In the famous sentence that begins the Gospel of St John all its rich fuzziness is put to use: 'In the beginning was the Word (Logos), and the Word was with God, and the Word was God'. Logos can mean a word, a thought or a system of thought, or a fundamental principle, and means them all here.

English words that end with *-logy* draw on the range of logos, which is greater than the range of 'logic', which signifies an art of logos, not its full extent. For instance 'geology' is the comprehensive and systematic study of ge, the earth, which will employ a kind of 'logic' as an instrument amongst others. So it is reasonable to ask: what logos, what -logy, does fuzzy logic belong to?

Faced with such a question, one answer would be to point to the gap and propose one obvious way to fill it: by coining the name 'fuzziology' as we have done, to refer to the body of knowledge of which fuzzy logic relates to. In this interpretation, fuzzy logic would be a technique (a word which itself comes from *techne*) used for understanding fuzziness. *Fuzziology then will be the comprehensive and systematic study of fuzziness itself.* In broad outline it should include the ontology (the theory of existence), epistemology (the theory of knowledge), praxiology (fuzziology as a theory of practice) and axiology (theory of value) on which fuzziology rests, which its techniques take for granted: the nature of the world it implies, the mind which knows this world, the techniques it uses and the values which drive fuzzy knowers to understand and act in such a world.

Fuzzy logic superficially sounds similar to fuzziology, but it is different in significant ways that are signalled by the form of the words themselves. In 'fuzzy logic' the two terms are separate, an adjective and a noun which together form a paradox (for most people). How can there be a 'logic' (which implies rigorous, clear thinking) and 'fuzziness' (which implies the opposite)? Fuzzy logic is a paradox, in which the two terms are put together but do not blend into a single concept. In 'fuzziology', in contrast, the two words have fused more completely, to form a new concept which is broader and less obviously paradoxical. It includes the same paradox, but in a form in which the difference has been transcended.

7.3 Transformations

The fuzzy thinking seems to make life complicated. We use and need ir because there is something about the world that asks for the fuzzy reasoning to describe them, some times, many times.

For instance, things change. Before they change they are one thing. Afterwards they are another thing, or the same thing but different. We need to see them fuzzy in the middle, between before and after. And since they never stop changing, they are fuzzy before 'before', and after 'after'.

'Transformations' are strong changes. The word comes from Latin *trans*, across, and *forma*, a form or shape: one shape or form crosses over to another. A becomes B. Before the process A and B may seem fixed and clear. During the 'becomes', A seems both A and not-A, and B seems B and not-B (though whatever it is, is not 'fuzzy' at that moment: it is just not easy to classify it as clearly A or B).

D'arcy Thompson, one of the founders of biomathematics, was fascinated with transformations in nature (Thompson, 1992). Writing in 1917 he used Cartesean co-ordinates to track the changes in various forms from one to another: for instance he showed that the face of a dog with only a few transformational tweaks can morph into a monkey, which can morph into a human face. With a few more adjustments, the face of a crocodile can morph into a dog, and thus into a human. The connections of course are not surprising in biology after Darwin. What his diagram shows is the impossibility of mapping the gap between the forms. If the forms are represented clearly, the transformation can be seen only through its effects on the different forms. If it is represented at all it could only be through a 'monster', a form that is moving between the two forms: a half-and-half form, a fuzzy form.

Much earlier than Darcy Thompson, with a more radical hypothesis, was the French biologist Etienne Geoffroy Saint Hilaire (1772 1844) who prepared the ground for Darwin by proposing, between 1800 and 1840, his doctrine of philosophical morphology. He demonstrated how the basic architecture of fish could be mapped onto the architecture of birds and mammals, linked through transformations, an idea Darwin drew on to argue evolutionary links. Geoffroy also argued, to the derision of most of his colleagues and the rest of the world of biology for the next 180 years, that the basic architecture of arthropods (insects, animals without backbones) and vertebrates (animals with skeletons) could be linked by a relatively small number of massive transformations (e.g., the external skeleton of the arthropod going inside to become the vertebrate skeleton). The derision stopped only in the 1990s, when genetic engineering found the genes that could do just that: turn the external skeleton of the arthropod into the internal skeleton of the vertebrate, or vice versa: like Kafka's fantastic story Metamorphosis, where a human woke up one morning as an insect.

This kind of transformation is difficult to imagine in stages. Geoffroy did not suppose that one fine morning at the dawn of time an insect gave birth to a vertebrate. He envisaged a kind of archetypal form, a common architectural resource that could be realised as either an arthropod (in many forms) or a vertebrate (again, in many forms).

This form is not simply an intermediate form, as in the Cartesian transformation of D'arcy Thompson. It is a deep form, so deep that it may not ever have existed in reality. How could anything live that had its skeleton half in and half out? Like the Cartesian intermediate form it is a 'monster', neither one thing or another, not A or B. But it is also not any intermediate thing either, so that it is neither not-A nor not-B, as well as being A and B to a degree. It is not so much intermediate as virtual. It is 'deeply' fuzzy: that is, it is hard to represent it in terms of categories that are more or less. They may be entirely different, yet systematically related.

Such a form *transcends* both the clear forms and even the fuzzy intermediate forms.

7.4 Transcendence

'Transcend' is an interesting word. *Trans* is the same root as in 'transform', across, or over. *Scend* comes from the Latin *scando*, to climb, from an Indo-European root *skan*, to leap. *Scando* combines with other prefixes, *ad* (towards) to ascend, and *de* (downwards) to descend. The English words have the same meanings. The second word suggests the root *skan* has a meaning of energy and a preference for movement upwards, but also includes movement downwards. That is, it contains the potential for a contradictory movement, up or down, just as Geoffroy's archetype had the potential to be arthropod or vertebrate.

This double potential can be seen in the Greek word *skandalon*, from which our 'scandal' derives. A *skandalon* was a trap, a stumbling block, something that made someone fall down, not leap up. Yet *skandalon* comes from the same word *skan*, which contained within itself the two possibilities, movement up and down.

'Transcendence' still carries the potential for the two movements, up and down. We think of 'transcendence' as passing over and above, in some higher realm, but the Geoffroy example is transcendence as a movement below, a 'deep' transcendence, into an earlier, 'lower' form which is the key to the transformation. It is 'deeply' (or 'highly') fuzzy: so monstrous that it is unimaginable, even in fuzzy terms.

Biologists did not in fact need to go back 500 million years to discover a reason to propose 'deeply' fuzzy or virtual forms. Behind the shelter of a cocoon a grub becomes a butterfly, in a transformational process that is clearly very deep. This is another type of ‘butterfly effect’: the reason why butterflies have been worshipped as a symbol of reincarnation by cultures as far afield as the Aztecs of Mexico and the ancient Greeks. Between life and death there is a similar transformation, a gap so great it seems impossible to leap under it or above it. But between life and death, a living person and a dead one, may be a 'deeply' fuzzy form, a monster that cannot be seen or described in clear or even fuzzy terms, whose symbol is a butterfly.

‘Man’ and ‘woman’ can be defined clearly, to differentiate them as much as possible with non-fuzzy categories. But men and women share almost all the same genes, their genetic character as humans, making men and women much more alike than different, becoming almost overlapping categories in spite of what the sexists say, with only a fuzzy boundary between them. Almost all that distinguishes them is their genitals, which seem so different as to be the biological basis for the difference, beyond argument so it seems.

But even that difference is less than it appears, or more superficial. The sexual organs seem to be a series of modifications of a common set of potential organs, so that what becomes a man’s penis is composed of what is the vagina and other sexual parts of a woman, the scrotum corresponds to the labia, and so on: a series of transformations as unbelievable as Geoffroy’s arthropods and vertebrates, producing forms so different that they seem unrelated. But it happened in very recent times for us all, in our first few months.

In a few cases the separate development into male or female does not happen as it should, and hermaphrodites are produced, with ‘fuzzy’ sexuality which is usually ‘monstrous’, not viable. But much more important, and nearly universal, is the common archetypal structure of sexuality which men’s and women’s sexuality draws on alike, which ‘transcends’ their sexual difference and underlies it, so deep they are normally not aware of it. This transcendent sexuality does not consist of a fuzzy lack of differentiation between men and women, a blurring of boundaries. On the contrary, its original form can only be projected as a virtual form by recognising the surface differences and tracing their intricate relationships, as a kind of dance around the mysterious transcendental form of a sex-which-is-not-one.

To illustrate the difference between 'transcendent' fuzzy categories and ordinarily fuzzy ones, we take the classic category of 'tall'. The two authors of this book are both more or less tall, but Vlad is the taller. As a transcendent category we take the fact that Vlad is Bulgarian, and Bob is Anglo-Celtic. These two categories seem entirely distinct from one another. However, Bulgaria 2500 years ago was populated by Thracian tribes, nomadic people with a culture that was similar to that of the Celts, who also lived there in that time as nomadic invaders. At the same time the Celts entered Ireland, Scotland and Wales, where Bob's Celtic ancestors ultimately came from. So Vlad and Bob share biological and cultural ancestors. This was 2500 years ago, so long ago it may seem irrelevant. But this is much less than the 500 billion years between modern arthropods and vertebrates and the moment of origin of their common architecture. And in spite of this apparent remoteness, it is a far more powerful force for creating a sense of identity between Vlad and Bob than their height, even if Bob was enough centimetres taller to be exactly as tall as Vlad.

The paradox of this form of fuzzy reasoning is that it uses the core belief of racism (that genetic differences are important in constructing identities and relationships) in order to transcend them. There is no human on earth with whom all other humans do not have a deep fuzzy similarity, because ultimately (and not very long ago compared to the splitting of vertebrates and arthropods) the human race is one. The same is true with differences only of degree not kind (that is, with only fuzzy differences) with every living thing, and even non-living things.

7.5 Understanding

In English, 'understanding' seems somehow 'deeper' than 'knowing', and this is built into the very form of the word: under - standing, standing under what ever it is that is understood. This probably seems so odd to many people they don't take it seriously as the wisdom carried by the word. In the frantic world of modern life we like to feel 'on top of' a problem, to take an 'overview'. But this word is telling us that if we wish to 'understand' we will not try to get outside and on top, but underneath.

There are other words that give the same message. For instance, 'hypothesise' comes from the Greek *hypo*, under, *thesis*, something that is put. 'Suppose' comes from the Latin equivalents: *sub* under, *ponere* to put. The hypothesis in science is the crucial step of creating a virtual world which will contain the solution to the problem is the world that is perceived. A hypothesis is not the same as understanding, because it is an attempt to be in that deep space, not an actual being there, as is the case with understanding. But hypotheses can be seen as a purposeful movement towards the deep space of 'understanding'.

7.6 The Social

The first part of 'social fuzziology' seems simple enough, in no need of further discussion. But even a common word like 'social' has a range of meanings, from 'friendly companionship' (as in a 'social club', a 'social life' to an organised group, usually of humans, divided horizontally into divisions based on inequalities of wealth, birth, power ('social class', 'social problem').

The word comes from Latin *socius*, which meant a friend or companion, and could also refer to a business partner or political ally. These are all kinds of relationship based on choice, free will, driven by desire, pleasure or profit. It is related to a root *sec-*, to follow, so that a *socius* is someone who follows you or whom you follow. *Societas* had a similar range of meanings, an association based on choice or desire. *Socius* was different from *civis*, a citizen, who was a member of a *civitatis*, a city or community united by political rights and obligations.

The path back into Indo-European from there is not clear, but there is a suggestive similarity to another cluster of Latin words around *sag-*, to perceive keenly, with the senses or intellect. The English 'seek' comes from the same Indo-European root, via the Old Saxon *sokian*, also to seek. *Sag-* also underlies *sag-us*, a sage, a wise person or prophet, who in the feminine could also be a procuress. The same root underlies the Greek *heg-*, to seek the way as guide, or to lead. This becomes the source of 'hegemony', leadership, the word used by the Italian Marxist Antonio Gramsci (1891 - 1937) to refer to the leading ideas of a society which are held in common as taken-for-granted truths but which also serve the interests of the ruling classes.

This history explains the fissure that constitutes the field of 'social'. It is still opposed to 'political' or 'civil', still containing a sense it is a group held together by the complex dynamics of reciprocal will and desire, though it also carries traces of a 2000 year history of appropriation of the 'social' by interests of power or profit. At the centre of its double meaning is the figure of the *socius*, individuals related by desire to other individuals, one following the other, each seeking the other. The 'society' formed by such *socii* is a network of reciprocal interactions formed by feelings and purposes in common, a form of relationship that is more fluid than all the alternative words (city, nation, state) though it is also part of the political process, as the origin of the word 'hegemony' reminds us.

'Social' therefore carries within it a concern for both power (manipulation, exploitation, appropriation) acting vertically, and love (association, comradeship, fellowship) acting horizontally, the difference between these two primary forces itself fuzzy, their continual interaction producing the endless fuzziness that over millions of years has made human society such a painful and desirable thing.

7.7 Conclusion

This chapter has been exclusively concerned with words, with ancient histories at the service of contemporary realities. That may make it seem irrelevant to the concerns of scientists and engineers in the 21th century, and indeed it is true that since the English Royal Society attempted to reform the language of science in the middle of the 17th century, science has been characterised by a militant distrust of natural language, and a desire for clarity at all costs. However, that is not the approach or fuzziology, which has consistently attempted to work with not against the principles of human language and the fuzziness of thought it encodes. As this chapter shows, some fuzzy key terms have a history that stretches back 8000 years, to a stage of society normally regarded as primitive and pre-scientific, yet that does not necessarily make them less complex or relevant. Since this history exists whether it is known or not, woven into the very fabric of our contemporary language and thought, it is better to know about it consciously. But there are also positive reasons to take an interest in this kind of study. To recover the lost history of the word 'fuzzy' is at the same time to connect fuzziology to a rich tradition of human thought, and give it access to its own complex genealogy. Terms like 'complexity', 'crisp' and 'edge' have a good feel to them. Behind that good feel is a conceptual wealth which fuzziology offers as a natural resource to everyone working in science or engineering as in business, politics or life.

References

Allen, J., Nelson, M. (1986) *Space Biopheres*, Synergestic Press

Arrow, K (1963) *Social Choice and Individual Values*, Yale Univ.

Bak, P. (1996) *How Nature Works*, NY: Copernicus

Barnev, P., Dimitrov, V. and Stanchev , P. (1974) Fuzzy System Approach to Decision-Making Based on Public Opinion Investigation through Questionnaires, *Proc. IFAC Symposium on Stochastic Control*, Budapest, Hungary

Barnev, P. and Dimitrov, V. (1975) Application of Fuzzy Sets Apparatus for Studying Artificial Languagues, in *Use of Mathematical Models and Computers in Linguistics*, Bulgarian Academy of Sciences: Sofia - in Russian

Bernal, M. (1987) *Black Athena : The Afroasiatic Roots of Classical Civilization*, Rutgers University Press

Blyth, R. (1983) *Zen and Zen Classics*, vol.3, Heian Intl Pub Co.

Bohm, D. (1980) *Wholeness and the Implicate Order*, Routledge Publ

Checkland P. 1981 *Systems Thinking, Systems Practice*, John Willey

Dimitrov, V. (1970) Generator of Heuristics in Human-Operator Models, in *Technical Cybernetics*, 2 (Series of the Institute of Cybernetics of Ukrainian Academy of Sciences: Kiev) - in Russian

Dimitrov, V. (1970) Algorithms of GMDH on Zadeh Fuzzy Sets, *Avtomatika*, 4 - in Ukrainian

Dimitrov, V. and Wechler. W. (1975) Optimal Fuzzy Control of Humanistic Systems, *Proc. 6th World Congress of IFAC*, Boston

Dimitrov, V., Driankov, D. and Petrov, A. (1975) Fuzzy Algorithms Execution, in *Use of Mathematical Models and Computers in Linguistics*, Bulgarian Academy of Sciences: Sofia - in Russian

Dimitrov, V. (1976) Non-Formal Theory of Fuzzy Management, *Systems and Control*, 1 - in Bulgarian

Dimitrov, V. (1976a) Social Choice and Self-Organisation under Fuzzy

Management, *Kybernetes*, 6, pp. 153-156

Dimitrov, V. (1976b) Learning for Decision-Making with Fuzzy Automata, in *Computer Oriented Learning Processes*, Ed. J. Simon, Noordhoff Leyden, pp. 149-153

Dimitrov, V. and Angelkov, V. (1976) Applications of Theory of Fuzzy Control in Economics, *Systems and Control*, 2 - in Bulgarian

Dimitrov, V., Stanchev, P. and Driankov, D. (1976) Applications of Theory of Fuzzy Control in Sociology and Medical Diagnostics, *Systems and Control*, 3 - in Bulgarian

Dimitrov, V. and Galabov, B. (1976) Using Fuzzy Sets in Processing of Empirical Information, *Proc. 2nd National Conference on Automation*, Varna, Bulgaria - in Bulgarian

Dimitrov, V. (1977) Modelling Humanistic Systems through Fuzzy Logic, *Proc. IFAC Symposium on Discrete Systems*, Dresden, Germany

Dimitrov , V. (1977a) Precise Effects through Fuzzy Actions, in *Applications and Research in Information Systems and Sciences*, vol.2, Hemisphere: Washington, London

Dimitrov, V. and Kunchev, O. (1977b) Efficient Governing Humanistic Systems by Fuzzy Instructions, in *Modern Trends in Cybernetics and Systems*, Springer Verlag: Heidelberg, New York, pp. 125-133

Dimitrov, V. and Driankova, V. (1977) Program System for Social Choice under Fuzzy Managing, in *Information Processing'77*, North Holland: Amsterdam, New York, pp. 807-811

Dimitrov, V. and Driankov, D. (1977) Pointwise Prediction Based on Fuzzy Sets, in Applications and Research in Information Systems and Sciences: vol.1, Hemisphere: Washington

Dimitrov, V., Driankov, D. and Petrov, A. (1977) Fuzzy Equations with Applications to Water Pollution Control, *Proc. IFAC Symposium on Environmental Planning, Design and Control*, 1977, Kyoto, Japan

Dimitrov, V. (1978) Fuzzy Control, in W. Wechler: *The Concept of Fuzziness in Automata and Language Theory*, Akademie-Verlag: Berlin, pp. 127-133

Dimitrov, V. (1979) Non-formal Theory of Fuzzy Management with Applications, *Doctor of Science Dissertation*, Russian Academy of Sciences, Moscow -

in Russian

Dimitrov, V. (1983) Group Choice and Self-Organisation under Fuzzy Information, *Fuzzy Sets and Systems*, vol.9, pp. 25-39

Dimitrov, V. (1983a) Fuzzy Catastrophes in Non-Trivial Decision-Making, in *Theory and Practice of Multiple Criteria Decision-Making*, Eds. C. Carlsson and Yu. Kochetkov, North Holland: Amsterdam, pp. 51-59

Dimitrov, V. (1989) Fuzzy Catastrophes, in *Systems & Control Encyclopedia*, Ed. M. Singh, Pergamon Press: Oxford, New York, pp. 1786-1788

Dimitrov, V. (1989a) On Artificial Fuzziness in Modelling, Proc. 3rd National Conference on *Methological Problems of Mathematical Modelling*, vol.3, Veliko Tarnovo, Bulgaria - in Bulgarian

Dimitrov, V. (1994) Fuzzy Symplectic Systems: A New Framework for Multi-Stakeholder Descision-Making, in *Applications of Fuzzy Logic in Complex Systems*: Proc. NAFIPS/IFIPS/ NASA International Joint Conference, San Antonio, Texas

Dimitrov, V. and Dimitrov, J. (1994) Fuzzy Logic: A Key to Shared Wisdom, in *Seized by Agreement, Swamped by Understanding: A collection of papers to celebrate the visit to Australia in August 1994*, by Humberto Maturana, Eds: L.Fell, D. Russell, and A. Stewart, University of Western Sydney, Hawkesbury, pp. 193-197

Dimitrov, J. and Dimitrov, V. (Eds.) (1996) *Fuzzy Logic and The Management of Complexity*, Richmond, NSW: FLAMOC'96

Dimitrov, V., Woog R. and Kuhn, L. (1996) The Divergence Syndrome in Social Systems, in *Complex Systems 96*, Eds. R. Stocker, H. Jelinek, B. Durnota and T. Bossomaier, IOS Press: Amsterdam, Oxford, Tokio, Washington, pp.142-146

Dimitrov, V. (1997) Use of Fuzzy Logic in Dealing with Social Complexity, *Complexity International*, vol.4 http://www.csu.edu.au/ci/vol04/dimitrov1/dimitrov.htm

Dimitrov, V. and Maheshwari, B. (1997) Fuzzy-Genetic Study of Hydraulic Resistance Families in Surface Irrigation, *Complexity International*, vol. 4, http://www.csu.edu.au/ci/vol04/dimmah/fuzzgen.htm

Dimitrov, V. and Kopra, K. (1998) Fuzzy Logic and the Management of Social

Complexity, in *Fuzzy Systems Design: Social and Engineering Applications,* Eds. L. Reznik, V. Dimitrov and J. Kacprzyk, Heidelberg, New York: Physica-Verlag, pp. 105-116

Dimitrov, V. (1999) Fuzzy Logic at Service for a Better World: The Social Dimension of Fuzzy Sets, in *Computational Intelligence and Applications*, Ed. N. Mastorakis, World Scientific Engineering Society Press: New York, Athens, pp. 129-132

Dimitrov, V. and Hodge, B. (1999) Virtual Meaning: Problems of Interpretation in the Social Sciences, in *Computing with Words in Information/Intelligent Systems*, Eds. L. Zadeh and J. Kacprzyk, Heidelberg, New York: Physica-Verlag, pp. 136-153

Dimitrov, V. and Ebsary, R. (2000) A Busca de Identidade, *Thot*, 74, pp. 51-60 (in Portugese); English version: http://www.pnc.com.au/~lfell/vlad2.html

Dimitrov, V. and Woog, R. (2000) Use of Fuzziology for Studying Global Economy, in *New Logic for the New Economy*, Ed. G. Zollo, Edizioni Scientifiche Italiane, Napoli, Italy, pp. 313-317

Dimitrov, V. and Korotkich, V. (Eds.) (2000) *With Fuzzy Logic in the New Millennium*, Richmond, NSW: Publ. UWS

Dimitrov, V. *et al* (2001) Fuzziology and Social Complexity, in "*Advances in Fuzzy Systems and Evolutionary Computation*", Ed. N. Mastorakis, World Scientific Engineering Society Press: New York, Athens, pp. 88-93

Dimitrov, V. and Korokich, V. (Eds.) (2002) *Fuzzy Logic: A Paradigm for the New Millennium*, Heidelberg, New York: Physica-Verlag

Dimitrov, V. and Hodge, B. (2002) Why Fuzzy Logic Needs the Challenge of Social Complexity? *With Fuzzy Logic in the New Millennium*, (Eds. V. Dimitrov and V. Korotkich), Heidelberg, New York: Physica Verlag

Dimitrov, V. (2002) Introduction to Fuzziology, in *Fuzzy Logic: A Framework for the New Millennium*, (Eds. V. Dimitrov and V. Korotkich), Heidelberg, New York: Physica Verlag

Dimitrov , V. and Weinstein, A. (2002) Social Fuzziology in Action: Dealing with Hard Negotiation, in *Fuzzy Logic: A Framework for the New Millennium*, (Eds. V. Dimitrov and V. Korotkich), Heidelberg, New York: Physica Verlag

Dimitov, V. and Woog, R. (2002) Fuzziology in Search of Insights for Understanding Global Economy, in *Fuzzy Logic: A Framework for the New Millennium*, (Eds. V. Dimitrov and V. Korotkich), Heidelberg, New York: Physica Verlag

Dimitrov, V. and Naess, T. (2001) Health through the Prism of Complexity, *Internet publication*, http://www.uws.edu.au/vip/dimitrov/health-and-complexity.htm

Dimitrov, V. (2001) Exploration of Complex Dynamics of Human Identity: Methodological Aspects, in "*Social Science Methodology in the New Millennium (Proceedings of the Fifth International Conference on Logic and Methodology, Cologne 3-6, 2000)*"; Joerg Blasius, Edith de Leeuw, Joop Hox and Peter Schmidt (eds.), on CD, Leske + Budrich, Opladen, Germany.

Dostoevsky, F. (2000) *Demons*, London: Everyman's Library/Random House, pp.403

Editorial (2001) *Science* 18 May, vol 292, p 1261

Feuerstein, K. (1990) *The Yoga-Sutra of Patanjali : A New Translation and Commentary*, Rochester, VT: Inner Traditions Intl Ltd

Gardner, H. (1993) *Frames of Mind: The Theory of Multiple Intelligences*, Basic Books

Greene, B. (2000) *The Elegant Universe: Superstrings, Hidden Dimensions, and the Quest for the Ultimate Theory*, Vintage Book

Grossberg, Stephen, ed. (1988) *Neural Networks and Natural Intelligence*, MIT Press, Cambridge, Mass

Goerner, S. (1994) *Chaos and the Evolving Ecological Universe*, Routledge

Hardt, M. and Negri, A. (2001) *Empire*, Harvard University Press

Heisenberg, W. (2000) *Physics & Philosophy: The Revolution in Modern Science*, Penguin

Herman, E. and Chomsly, N. (2002) *Manufacturing Consent : 'The Political Economy of the Mass Media*, Pantheon Books

Horgan, J. (1996) *The End of Science*, Perseus Pr.

Huxley, A. (1990) *The Perennial Philosophy*, HarperCollins

John, F. (2001) *An Elementary Textbook of Ayurveda: Medicine With a Six Thousand Year Old Tradition*, Psychosocial Press

Judith, A. (1987) *Wheels of Life: A User's Guide to the Chakra System*, Llewellyn

Kauffman L. (1997) Virtual Logic - Fixed Points and Paradoxes, *Cybernetics and Human Knowing*, 4, p.65

Kosko, B. (1995) *Fuzzy Thinking*, HarperCollins Publishers

Kirkegaard, S. (1958) *Philosophical Fragments*, Princeton University Press

Krishnamurti, J. (1974) *The First and Last Freedom*, New York: Harper and Row

Langton, C. (1989) Artificial Life, in *Artificial Life, vol.VI of SFI Studies in the Science of Complexity* (ed. C. Langton), Redwood City, CA: Addison-Wesley

Lawlor, R. (1991) *Voices of the First Day*, Vermont: Inner Traditions Inren. Ltd.

Lehar, S. (1999) Harmonic Resonance Theory: an Alternative to the "Neuron Doctrine", *Internet publication*, http://cnsalumni.bu.edu/pub/slehar/webstuff/hr/hr.html

Liddell, H. and Scott, R. (1849) *A Greek-English Lexicon* Oxford: Oxford University Press

Lorenz, E. (1993) *The Essence of Chaos* Washington: University of Washington Press

Mandelbrot, M. (1982) *The Fractal Geometry of Nature*, San Francisco: Freemen Co.

McNeil D. and Dimitrov, V. (1998) On the Topolology of Uncertainty, in *Fuzzy Systems Design: Social and Engineering Applications,* Eds. L. Reznik, V. Dimitrov and J. Kacprzyk, Heidelberg, New York: Physica-Verlag, pp. 171-184

Mitchell, S. (1992) *Tao Te Ching : A New English Version* , Harper Perennial

Morson, G.S. and Emerson, C. (eds) (1989) *Rethinking Bakhtin: Extensions and Challenges*, Evanston, Ill.: Northwestern Univ. Press

Nelson, R. (1999) The Global Consciousness Project: Consciousness Resonance, *Internet publication*, http://noosphere.princeton.edu/resonance.html

Nietzsche, F. (1956) *The Birth of Tragedy*, transl. by W. Kaufmann, NY:

Doubleday

Pantanjai, B (1993) *Yoga Sutras of Patanjali*, Brotherhood of Life Books

Partridge, E. (1966) *Origins* London: Routledge and Kegan Paul

Price, H., (1996) Time's Arrow and Archimedes' Point, New York: Oxford University Press

Prigigine, I. and Stengers, I. (1984) The End of Certainty : Time, Chaos, and the New Laws of Nature, Free Press

Ramonet, I (2001) An Enemy. At last, *Le Mond Diplomatique*, October, http://www.en.monde-diplomatique.fr/2001/10/0

Reznik, L., Dimitrov, V. and Kacprzyk, J. (Eds.) (1998) *Fuzzy Systems design: Social and Engineering Applications*, Heidelberg, New York: Physica-Verlag 1

Riley, M. (2001) *Stdney Morning Herald*, Friday, August 3, p.1

Russell, D, Dimitrov, V. and Fell, L. (1994) Ancient Wisdom and Contemporary Ecological Problems in *Seized by Agreement, Swamped by Understanding: A collection of papers to celebrate the visit to Australia, in August 1994, by Humberto Maturana*, Eds: L.Fell, D. Russell, and A. Stewart, University of Western Sydney, Hawkesbury, pp. 207-215

Saussure, F de (1974) *A Course in General Linguistics* trans W Baskin, London: Fontana

Steiner, R. (1998) *Art as Spiritual Activity*, Hudson, NY: Anthroposophic Press

Thompson, D. (1992) *On Growth and Form*, NY: Dover Publ.

Taylor, C. et all (eds.) (2001) *Greek Philosophers: Socrates, Plato, Aristotle*, Oxford Univ. Press

Van Eeenwyk, J. (1997) *Archetypes and Strange Attractors*, Toronto: Inner City Books

Vygotsky, L (1987) *The Collected Works of L.S. Vygotsky. Vol.1: Problems of General Psychology*, Eds. R.W. Rieber and A.S. Carton, in translation of N. Minick, NY: Plenum Press

Waldrop, M. (1992) *Complexity: the Emerging Science At the Edge of Order and Chaos*, Touchstone Books

Wechler, W. and Dimitrov, V. (1974) R-fuzzy Automata, in *Information Processing'74*, North Holland: Amstrdam, New York, pp. 657-660

Westphal, M. (1996) *Becoming a Self: A Reading of Kierkegaard's Concluding Unscientific Postscript*, Purdue University Press

White, J. and Riddle, J. (1876) *A latin-English Dictionary.* London: Longman, Green and Co.

Williams, R. (1988) *Keywords: A Vocabulary of Culture and Sociaty*, London: Fontana

Wittgenstein, L. (1994) Philosophical Investigations, Oxford: Blackwell

Woog, R., Dimitrov V. and Kuhn-White, L. (1998) Fuzzy Logic as an Evocative Framework for Studying Social Systems, in *Fuzzy Systems Design: Social and Engineering Applications,* Eds. L. Reznik, V. Dimitrov and J. Kacprzyk, Heidelberg, New York: Physica-Verlag, pp. 105-116

Woog, R., Kuhn, L. and Dimitrov, V. (1996) Not Just Going to the Abyss, Bridging It Towards the Articulation of a Spiritual-Critical Heuristic Inquiry Methodology, in *Fuzzy Logic and the Management of Complexity: Proceedings of the First International Discourse on Fuzzy Logic and the Management of Complexity*, Eds. J.Dimitrov and V. Dimitrov, Richmond, NSW: FLAMOC'96, pp. 13-17

Zadeh, L. (1965) Fuzzy Sets, *Information and Control*, 8, pp.338-359

Zadeh, L. (1973) A New Approach to Analysis of Complex Systems, *IEEE Trans. Syst., Man, Cybern,*, SMC-3, 1

Zadeh, L. (1999) What is Computing with Words? in *Computing with Words*, eds. L. Zadeh and J. Kasprzyk, Heodelberg and new York: Physica-Verlag

GPSR Compliance

The European Union's (EU) General Product Safety Regulation (GPSR) is a set of rules that requires consumer products to be safe and our obligations to ensure this.

If you have any concerns about our products, you can contact us on ProductSafety@springernature.com

In case Publisher is established outside the EU, the EU authorized representative is:

Springer Nature Customer Service Center GmbH
Europaplatz 3
69115 Heidelberg, Germany

Batch number: 10725802

Printed by Printforce, the Netherlands